Climate-Resistant Smart Agriculture for Healthy Food Production

Eduard Babulak
https://orcid.org/0000-0003-2450-090X
National Science Foundation, USA

IGI Global
Scientific Publishing
Publishing Tomorrow's Research Today

Vice President of Editorial	Melissa Wagner
Director of Acquisitions	Mikaela Felty
Director of Book Development	Jocelynn Hessler
Production Manager	Mike Brehm
Cover Design	Jose Rosado

Published in the United States of America by
IGI Global Scientific Publishing
701 East Chocolate Avenue
Hershey, PA, 17033, USA
Tel: 717-533-8845 | Fax: 717-533-7115
Website: https://www.igi-global.com E-mail: cust@igi-global.com

Library of Congress Cataloging-in-Publication Data

Names: Babulak, Eduard, 1957- editor
http://id.loc.gov/vocabulary/relators/edt
http://id.loc.gov/authorities/names/n2023007013
http://id.loc.gov/rwo/agents/n2023007013
Title: Climate-resistant smart agriculture for healthy food production / edited by Eduard Babulak.
Description: Hershey, PA : IGI Global Scientific Publishing, [2026] | Includes bibliographical references and index. | Also available online.
Identifiers: LCCN 2025027604 (print) | LCCN 2025027605 (ebook) | ISBN 9798337348292 ebook | ISBN 9798337348278 hardcover | ISBN 9798337348285 paperback
Subjects: LCSH: Artificial intelligence
http://id.loc.gov/authorities/subjects/sh85008180 | Machine learning
http://id.loc.gov/authorities/subjects/sh85079324
Classification: LCC Q335 (ebook) | LCC Q335 .C54 2026 (print) | DDC 303.48/34 23/eng/20250--dc25
LC record available at https://lccn.loc.gov/2025027604
British Cataloguing in Publication Data
A Cataloguing in Publication record for this book is available from the British Library.

I would like to dedicate this Book to my family and friends who provided me with their kind support and encouragement during the Book project.

Eduard Babulak

2025-03-08

Editorial Advisory Board

Table of Contents

Detailed Table of Contents

of which are crucial given the world's expanding population. Although AI has enormous potential for the agriculture industry, there are many challenges (e.g., need for additional data, creation of accessible & reasonably priced AI solutions), & requirement for farmer education/training. Enhancing AI algorithms, creating fresh uses for AI in agriculture & resolving adoption barriers will be the main goals of future research/development.

P. Selvakumar, Department of Science and Humanities, Nehru Institute of Technology, Coimbatore, India
H. Vidyalakshmi, JSS Academy of Higher Education and Research, India
R. Josphineleela, Panimalar Engineering College, India
V. Vaissnave, SRM Institute of Science and Technology, India
T. Ragupathi, SRM Institute of Science and Technology, India
Raj Kumar, Maharshi Dayanand University, Rohtak, India
T. C. Manjunath, Rajarajeswari College of Engineering, India

In decades, technological innovation has profoundly reshaped industries across the globe, and agriculture—long rooted in traditional methods—is no exception. transformative powerful catalyst for change, driving a new era in food production that emphasizes efficiency, sustainability, and precision. As the global population steadily climbs toward by over 50%, (FAO). Meeting this demand in a context of finite arable land, climate volatility, labor shortages, and growing environmental concerns requires a rethinking of conventional agricultural practices. AI with learn from patterns, and make informed decisions, agricultural role in food production spans a wide spectrum—from crop selection and planting to harvesting, packaging, and distribution. This approach leverages tools like machine learning, optimize resource use, reduce waste, and determine, predict crop diseases, monitor weather conditions in real time.

Food adulteration is an ongoing and dynamically increasing challenge to public health, economic authenticity, and food system confidence around the world. This survey provides an overview of recent developments in the detection of food adulteration, with special emphasis on the utilization of Internet of Things (IoT) technologies and machine learning strategies for real-time inspection and analysis. The research classifies adulterants, impacted food products, and related health effects, as well as examines traditional and novel detection methods such as spectroscopy, chromatography, and sensor-based systems critically. Focus is given to IoT-based frameworks that employ an assortment of physical and chemical sensors for decentralized and autonomous adulteration identification. In addition, the chapter compares current implementations, limitations in sensor calibration, data accuracy, standardization, and the lack of region-specific datasets for machine learning model training. Through an outline of the present technological state and an enumeration of research priorities, including the interoperability, regulatory support, and data privacy, this research is targeted at future developments toward scalable, cost-efficient, and smart food safety solutions. The survey highlights the revolutionary role of smart technologies in restructuring food quality assurance and safeguarding consumer health in developed and developing settings.

This paper examines the development and implementation of climate-resistant smart agriculture policies in Taiwan and Indonesia, highlighting their contributions to healthy and safe food production as a strategic pathway toward enhanced food security and sustainable rural development. Both countries face increasing challenges from climate change, including extreme weather events, shifting crop patterns, and threats

Preface

The ongoing increase of the unpredictable and very powerful natural phenomena worldwide, the humanity survival depends on agriculture, healthy food production and distribution. As climate change continues to disrupt traditional agricultural practices, the integration of smart technologies has emerged as a solution for sustainable food systems. These cutting-edge innovations are transforming the way people cultivate and harvest food. This approach represents a vital step toward achieving global food security while safeguarding the health of both people and the planet.

In decades, technological innovation has profoundly reshaped industries across the globe, and agriculture long rooted in traditional methods is no exception. The transformative powerful catalyst for change is driving a new era in food production that emphasizes efficiency, sustainability, and precision. Climate-Resistant Smart Agriculture for Healthy Food Production explores the developments in the field of smart agriculture to help fight off climate change. This book presents the most effective technological and business solutions for the benefit of mankind. Covering topics such as agriculture, climate change, and smart technology, this book is an excellent resource for environmental scientists, agricultural professionals, policymakers, researchers, educators, academicians, and more.

The book attracted seventy nine (79) chapter proposals addressing important subject related issues. The AI is becoming more widely acknowledged as a potent instrument for transforming farming methods. A more sustainable & effective food production system is required due to the growing global population, & AI has various benefits in this area. The transformational potential of AI in agriculture is the main topic of this chapter, which also emphasizes how it may improve resource management, address food security, & raise overall agricultural output. A more sustainable & effective agricultural sector is facilitated by AI-powered systems that provide accurate solutions for crop monitoring, disease identification, resource optimization, & increased yield all of which are crucial given the world's expanding population. The book attracted lot of interests from scholars worldwide. The book brings to light how artificial intelligence (AI) supports technological development in the field of agriculture and healthy food production.

The current environmental disasters such as, drafts, flooding, forest fires, earthquakes, wars, air and water pollutions, the humanity needs guarantees for effective and productive agriculture, and healthy food production. Studies suggest the one of the greatest challenges for humanity will be productive agriculture and healthy food production. The book brings to light the current-state-of-the-art and future research, innovation and development in the field of smart agriculture and food production assisted by Artificial Intelligence.

The book main objective is to bring to light the state-of-the-art and future directions in research, innovation and development of next generation of the AI inspired Climate Resistant Smart Agriculture and Healthy Food Production. The book provides up-to-date information of the current transformation in agriculture and healthy food production, and brings clarity on the importance and benefits of applied AI in the field of Smart Agriculture and Smart Healthy Food Production. The book promotes creation of multidisciplinary teams of experts in the nation and worldwide.

The book will be most valuable resource for the Agriculture and Food Production Industry Decision Makers, Practitioners, Researchers, Innovators, Developers, University Faculty, Senior, Graduate and Post-Graduate students. The book will be important reference for decision makers, technical staff in the Agriculture and Food Industries, Government, and Business worldwide.

The Book presents concise introduction to smart technology driven agriculture, followed by the most recent developments in the AI Assisted Smart Agriculture and the AI Assisted Healthy Food Production. The book addresses important issues concerning the Future Smart Farming, Climate Resistant Agriculture, and Smart Food Production and Distribution. The

Conclusions and Future Research Directions highlight the current state of the art and future challenges and possible solutions for the years to come.

Years back, before having machine assisted agriculture, many dreamt about a day when they could get healthy food from the local store. Today, all this are almost forgotten memories of ancestors and our and younger generations today very seldom discuss or think about it. We all live in the age of the third millennium technological era where Smart Agriculture is driven by the AI and often administered by farmers and their family members.

Today, people expect to find high quality food products not worrying much about the agriculture which is often outside in the open country and far from the cities. Perhaps, tomorrow our grandchildren will discuss the Technological Advancements in the Smart Agriculture as part of Super-Intelligent AI driven Ultra-Smart Cyberspace. I humbly believe that we are yet at the very beginning of new era of ultra-smart agricultural technological revolution that will transform the way of Healthy Food Production and Distribution worldwide. I sincerely hope that you will find the book educational, practical and inspiring while looking for ways to explore and master what may be considered today as impossible and futuristic.

- An overview of the subject matter

The era of AI driven Smart Agriculture and Healthy Food Production is here. The applications of AI with current and future dynamic trends in research, innovation and developments of Smart Agriculture and Healthy Food Production will contribute to the betterment of mankind worldwide.

Given the current state of the art in the world of the Smart Agriculture, apart from technological advancements it is also very important to assess the AI impact on Health Food Production and Distribution today and tomorrow. The many academic areas covered in this publication include, but are not limited to:

- Artificial Intelligence (AI)
- Crop Detection
- Ethical Compliance

- Healthy Food Productions
- Leadership Styles
- Managerial Performances
- Research Productivity
- Safe Food Production
- Smart Agricultural
- Smart Farming
- Sustainable Organization
- A description of where your topic fits in the world today

In support of the AI technological advancements it is important to have a better understanding of the AI practical applications in current and future Smart Agriculture. The advancements of research in the field of Smart Climate Resistant Agriculture, combined with the research in the AI & Ultra-Smart Computing are essential to humanity survival. The AI driven Smart Agriculture and Healthy Food Production today, is utilizing Advanced Computational Systems while collecting, processing and analyzing a real-time to guarantee a good harvest despite the very complex and unpredictable natural phenomena.

Due Climate Change, the humanity lives in the most dramatic times, yet despite of its most negative impact it does also inspire dynamic innovation, research and developments in the world of education, health, business, government, industry, plus others. The ranges and scale of possible applications of AI in the Smart Agriculture and Healthy Food Protection has become ubiquitous and pervasive in societies of today and tomorrow worldwide.

- A description of the target audience

The book is primarily intended for the following type of audience.

1. Scholastic, Industry & Business Innovation and Development Research and Professional Communities worldwide: The book will be an essential reference for researchers in the subject domain having prerequisites of AI & Agriculture & Food Production & Distribution.

2. University Faculty: The book presents practical quality technical know-how in the relevant field to the undergraduate/post-graduate students.
3. Postgraduates & Senior Undergraduates: The book will provide excellent resources to pursue higher studies in the form of research inspired by the different facets of AI & Agriculture & Food Production & Distribution principles presented in book chapters with potential contributions to curriculum development.
4. Academic and corporate libraries: Adding this volume to libraries would enrich the content regarding the research base.
    - A description of the importance of each of the chapter submissions (this entails providing a paragraph description of **each** chapter)

CHAPTER OVERVIEW

The first chapter explores the integration of artificial intelligence (AI) into climate smart agriculture (CSA) to enhance resilience, sustainability, and food security amid climate change. It examines how AI technologies such as machine learning, computer vision, and predictive analytics support precision agriculture, early warning systems, soil and water management, pest control, and supply chain optimization. Drawing on case studies from India, Sudan, Kenya, Japan, the United States, and China, the chapter illustrates AI's role in transforming food systems. It also analyzes key challenges including limited data access, digital inequality, ethical concerns, and weak institutional frameworks. The chapter presents strategic recommendations for building inclusive digital ecosystems, promoting ethical AI design, and strengthening public private partnerships. By aligning AI innovations with sustainability objectives, the chapter positions AI as a critical enabler of climate resilient agriculture and global development efforts.

The second chapter illustrates how AI is becoming more widely acknowledged as a potent instrument for transforming farming methods. A more sustainable & effective food production system is required due to the growing global population, & AI has various benefits in this area. The transformational potential of AI in agriculture is the main topic of this chapter, which also emphasizes how it may improve resource management, address food security, & raise overall agricultural output. A more sustainable & effective agricultural sector is facilitated by AI-powered systems that provide accurate solutions for crop monitoring, disease identification, resource optimization, & increased yield—all of which are crucial given the world's expanding population. Although AI has enormous potential for the agriculture industry, there are many challenges (e.g., need for additional data, creation of accessible & reasonably priced AI solutions), & requirement for farmer education/training. Enhancing AI algorithms, creating fresh uses for AI in agriculture & resolving adoption barriers will be the main goals of future research/development.

The third chapter discusses a multiple transformations since its inception over 10,000 years ago. From the use of primitive tools in the Neolithic period to the mechanized farming practices of the 20th century, each revolution has drastically improved productivity and efficiency. Today, the world stands brink driven by. As diminishing arable land, and labor shortages threaten food security, to reshape modern and ensure sustainable food production for future generations. The authors discuss the integration of intelligent systems which are capable of mimicking human reasoning and decision making into the agricultural processes. These systems analyze vast datasets collected drones to generate insights, predictions, and recommendations that can guide farming decisions with unprecedented precision. From to pest control resource, AI is becoming a pivotal force in making agriculture more transformative impacts AI is seen in precision farming, an approach that tailors farming practices to specific field farmers can determine, predict needs, where necessary.

Chapter 4 shows evidence that food adulteration is a major risk to public health, usually resulting in severe health issues as well as financial loss. This research envisions an Internet of Things (IoT) smart system developed for real-time adulterant detection of frequently adulterated foods, facilitating healthy consumption and enabling healthier living. The suggested framework combines a sensor network of inexpensive sensors with microcontrollers for the measurement of physical and chemical parameters like pH, turbidity, conductivity, and color variation, which are signifying adulteration in products such as milk, spices, oil, and vegetables. The data received from these sensors are wirelessly transmitted to a cloud platform for storing, analyzing, and generating alerts. Machine learning algorithms are utilized for decision-making and pattern recognition to accurately classify adulterated samples. An interface for a mobile application gives users immediate alerts and extensive reports of adulteration. Experimental findings prove the success of the system in identifying general adulterants like starch in milk samples, metanil yellow in turmeric samples, and detergent in water samples. The solution is highly accurate, portable, and scalable and can be most appropriately implemented in household, retail, and regulatory sectors. This study helps to create a safer food environment by intelligent monitoring and gives power to the consumers to make informed decisions.

Chapter 5 examines the development and implementation of climate-resistant smart agriculture policies in Taiwan and Indonesia, highlighting their contributions to healthy and safe food production as a strategic pathway toward enhanced food security and sustainable rural development. Both countries face increasing challenges from climate change, including extreme weather events, shifting crop patterns, and threats to food systems. In response, Taiwan has adopted precision agriculture and digital technology integration through its "Smart Agriculture 4.0" initiative, while Indonesia is advancing its "Agricultural Digitization Roadmap" and climate-resilient farming programs. Through a comparative analysis, this study evaluates policy frameworks, institutional support, technological innovations, and grassroots engagement in each country. The findings indicate that the integration of Internet of Things (IoT), AI, and sustainable agro-ecological practices has significantly improved crop resilience, minimized chemical inputs, and enhanced rural livelihoods.

Chapter 6 investigates the crucial link between green and renewable energy-based financial literacy and its effects on promoting sustainable food sovereignty and security in México. The principal findings indicate that farmers who possess higher renewable energy financial literacy are more likely to adopt solar-powered irrigation systems, biomass energy services, and energy-efficient agricultural practices. This adoption not only reduces the cost of operations but also enhances climate change resilience and food production sustainability. The paper concludes that investments in targeted green energy financial literacy programs, coupled with revved-up financing mechanisms for renewable energy technologies in agriculture, can significantly advance México towards its goals of sustainable food security with environmental stewardship. These findings have important policy implications for the development of policy, educational curricula, and financial institution strategy that aim to facilitate México's agricultural transition.

Chapter 7 presents the book subject matter in concise manner. The chapter discusses how the integration of Artificial Intelligence (AI) in agriculture is reshaping how farmers adapt to climate change through Climate Smart Agriculture (CSA). CSA emphasizes resilience, environmental sustainability, and productivity. AI enhances these goals via real-time data collection, predictive modeling, and decision support systems. The chapter highlights AI's applications in precision farming, climate risk assessment, and resource management. Examples include AI-driven weather forecasting, smart irrigation, pest control, and livestock monitoring. These innovations significantly boost farming system resilience. However, challenges such as data accessibility, high costs, and ethical concerns persist. The chapter outlines strategies to scale AI in CSA, stressing the importance of robust data infrastructure, inclusive policies, and public-private partnerships. Ensuring affordable access to AI for smallholder farmers is crucial to achieving sustainable and climate-resilient agriculture globally.

- A conclusion of how your book impacts the field and contributes to the subject matter

The most recent advances in the Applied Artificial Intelligence in Smart Agriculture and Healthy Food Production contribute to creating a future Ultra-Smart Intelligent Agricultural Engineering Systems, which may be capable of monitoring, analyzing and predict the natural phenomena for the benefit of good harvest. These future Ultra-Smart Intelligent Agricultural Engineering Systems will be far beyond sophistication of a current Agricultural Systems. There are currently ongoing discussions concerning the potential value and impact of the AI assisted Ultra-Smart Agriculture on society today and tomorrow.

Applied AI and Ultra-Smart Agriculture has become an integrated part of curriculum in the Academia, and there are many applications in Government, Business and Industry worldwide. The current dynamic advancements if the field of AI Assisted Agriculture and healthy Food

Production will continue impacting all aspect of human lives for the years to come.

The Book main objective is to promote creation of *Global Center of Excellence in AI Assisted Smart Agriculture and Healthy Food Production* for research, innovation and development of the AI driven Ultra-Smart Intelligent Agricultural Engineering Systems.

The second objectives is to promote a better understanding and clarity of the value that the AI and Smart Agriculture present today and will present tomorrow to the world of the Ultra Smart Agriculture and Healthy Food Production and ultimately to the society worldwide.

The third objective is to establish the Assessment Metrics for the Ultra Smart Agriculture and Healthy Food Production Social Impact on Public and Scientific Communities worldwide.

The Book discusses the fundamental principles of Ethics and Professionalism, while providing assessment of the applied AI applications in Smart Agriculture and Healthy Food Production social impact on national and global levels.

It also promotes seamless creation of multidisciplinary teams of experts in the nation and worldwide. Reader may seek the answers to a question like: Will the AI be well understood and become part of our daily agriculture or else?

The Book brings clarity and better understanding of the value that AI and Smart Agriculture brings to society, which will ultimately open gate to new opportunities for future innovation, research and development in this field of studies.

Eduard Babulak

2025-15-08

Acknowledgment

I would like to use this opportunity to express my sincere gratitude to contributors, reviewers, to Ms Amelia Bucher, Mrs. Jocelynn Hessler, Ms Jordan Walter, Ms Casey Long, Ms Ms Eddinger, Ms Wagner, Ms Barrantes, Ms Travers, Ms McLouglin, Ms Wertman and the IGI Global Colleagues and Friends for their excellent support, kind guidance and good counsel during the course of this and my earlier projects. I would like to say great thank you to my family and friends for their kind support and encouragement during the Book project.

Eduard Babulak

2025-05-08

Introduction

Given the current ongoing climate change, environmental disasters such as, drafts, flooding's, forest fires, earthquakes, wars, air and water pollutions, the humanity needs guarantees for effective and productive agriculture, and healthy food production. Studies suggest the one of the greatest challenges for humanity will be productive agriculture and healthy food production. The book will bring to light the current-state-of-the-art and future research, innovation and development in the field of smart agriculture and food production assisted by Artificial Intelligence.

The book main objective is to bring to light the state-of-the-art and future directions in research, innovation and development of next generation of the AI inspired Climate Resistant Smart Agriculture and Healthy Food Production. The book will provide up-to-date information of the current transformation in agriculture and healthy food production, and bring clarity on the importance and benefits of applied AI in the field of Smart Agriculture and Smart Healthy Food Production. The book will promote creation of multidisciplinary teams of experts in the nation and worldwide.

The book will be most valuable resource for the Agriculture and Food Production Industry Decision Makers, Practitioners, Researchers, Innovators, Developers, University Faculty, Senior, Graduate and Post-Graduate students. The book will be important reference for decision makers, technical staff in the Government, Military, Business, Agriculture and Food Manufacturing Industry worldwide.

The Book presents concise Introduction to Technology Driven Agriculture, followed by the most recent developments in the AI Assisted Smart Agriculture and the Ai Assisted Healthy Food Production. The book addresses important issues concerning the Future Smart Farming, Climate

Resistant Agriculture, and Smart Food Production and Distribution. The Conclusions and Future Research Directions highlight the current state of the art and future challenges and possible solutions for the years to come.

Eduard Babulak

National Science Foundation, USA

Chapter 1
AI-Inspired Climate Resistant Smart Agriculture and Healthy Food Production

Asamoah Oppong Zadok
https://orcid.org/0009-0001-6778-6250
Lincoln University of Missouri, USA

ABSTRACT

This chapter explores the integration of artificial intelligence (AI) into climate smart agriculture (CSA) to enhance resilience, sustainability, and food security amid climate change. It examines how AI technologies such as machine learning, computer vision, and predictive analytics support precision agriculture, early warning systems, soil and water management, pest control, and supply chain optimization. Drawing on case studies from India, Sudan, Kenya, Japan, the United States, and China, the chapter illustrates AI's role in transforming food systems. It also analyzes key challenges including limited data access, digital inequality, ethical concerns, and weak institutional frameworks. This chapter, per the authors, presents strategic recommendations for building inclusive digital ecosystems, promoting ethical AI design, and strengthening public private

DOI: 10.4018/979-8-3373-4827-8.ch001

partnerships. By aligning AI innovations with sustainability objectives, the chapter positions AI as a critical enabler of climate resilient agriculture and global development efforts.

1. INTRODUCTION

Agriculture, the cornerstone of human civilization, is now at the center of one of the greatest challenges of the 21st century: climate change. As extreme weather events, erratic precipitation patterns, rising global temperatures, and deteriorating soil health increasingly disrupt global food production systems, the world faces a dual crisis of food insecurity and environmental degradation. According to Ofori, Cobbina, and Obiri (2021), climate variability and change already threaten food availability, access, and utilization, particularly in vulnerable regions such as sub-Saharan Africa, South Asia, and Latin America.

In response to these challenges, there is an urgent need to redesign agricultural systems to be climate-resilient, sustainable, and equitable. Climate-Smart Agriculture (CSA) offers a transformative framework. Introduced by the Food and Agriculture Organization (FAO), CSA integrates three core pillars: increasing productivity and incomes, adapting to and building resilience against climate change, and reducing greenhouse gas emissions where feasible (Sulaiman, Chuluunbaatar & Vishnu, 2018). Despite its potential, CSA implementation faces challenges, including limited access to data, weak institutional coordination, and insufficient farmer involvement.

Artificial Intelligence (AI) offers promising solutions to these limitations. AI, which includes machine learning, deep learning, computer vision, and natural language processing, has been applied in multiple sectors and is now influencing agriculture. It enables the analysis of large datasets, real-time predictions, automation of processes, and resource optimization (Kumar et al., 2023; Assimakopoulos et al., 2025).

When AI technologies are integrated with CSA principles, a new model emerges: AI-inspired climate-resistant smart agriculture. This model enhances the precision and efficiency of farming while also increasing adaptive capacity for climate resilience. It supports healthy food production by improving crop selection, minimizing the use of harmful chemicals, advancing food traceability, and supporting data-driven nutrition strategies (Bulut and Arslan, 2022; William et al, 2024).

AI technologies support farmers in predicting pest outbreaks, selecting drought-resistant crops, optimizing irrigation systems, and monitoring soil quality (Patil, 2024). Development agencies and governments use AI-powered tools for mapping climate risks and planning interventions. Consumers benefit from improved food safety and greater transparency. As digital infrastructure expands through Internet of Things (IoT), cloud computing, mobile applications, and blockchain systems, the integration of AI into agriculture becomes both necessary and achievable.

This chapter examines the theoretical foundation, practical applications, and potential of integrating AI into climate-smart agriculture. The following core questions are addressed:

- How can AI enhance the resilience, efficiency, and sustainability of agricultural systems under climate stress?
- What technological and social innovations are driving AI-based food systems?
- What barriers need to be overcome to scale these innovations?
- How can AI contribute to producing diverse, safe, and climate-resilient food?

The analysis draws on peer-reviewed research, global case studies, and reports from regions such as Sudan, India, and sub-Saharan Africa. Examples include smart irrigation, predictive pest models, digital supply chains, and soil health monitoring systems. The chapter also discusses challenges such as digital inequity, knowledge gaps, and policy fragmentation.

The goal is to demonstrate how AI-inspired agriculture can contribute to global food and climate goals, particularly the United Nations Sustainable Development Goals: SDG 2 (Zero Hunger), SDG 12 (Responsible Consumption and Production), and SDG 13 (Climate Action). As the world approaches a population of nearly 10 billion by 2050 under worsening climate conditions, integrating AI into agriculture is not just an option but an urgent necessity.

2. BACKGROUND AND RATIONALE

2.1 Climate Change and Agricultural Vulnerability

Climate change is rapidly altering the global agricultural landscape. Its effects are no longer hypothetical projections but realities felt daily by farmers across the globe. These effects include prolonged droughts, unexpected flooding, rising average temperatures, shifts in growing seasons, increased prevalence of pests and diseases, and severe land degradation (Kumar et al., 2018; Reddy & Reddy, 2015). These shifts disproportionately affect low- and middle-income countries, where agriculture contributes significantly to GDP and where food systems are heavily dependent on rainfall and fragile ecosystems, making them especially vulnerable to climate variability and shocks (Gautam et al., 2018). This multidimensional relationship between climate change and agriculture is illustrated in Figure 1, which depicts how environmental stressors such as heatwaves, droughts, and pest outbreaks cascade through food systems to impact productivity, resilience, and food security (Singh & Singh, 2017).

Figure 1. Interlinkages Between Climate Change and Agriculture

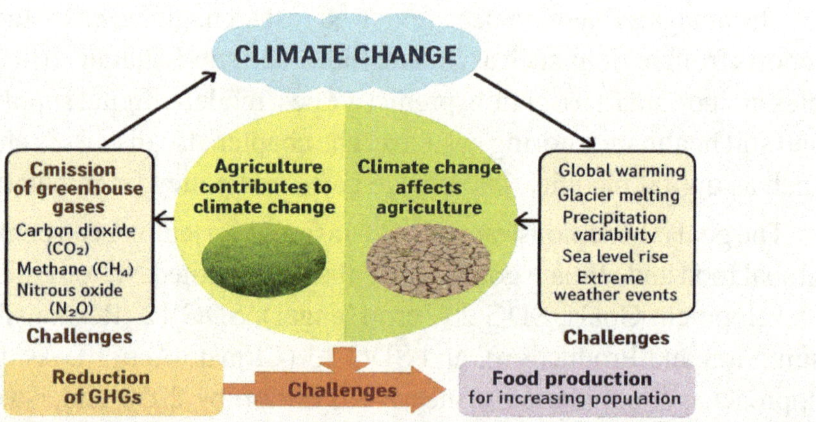

Source: Singh & Singh (2017)

The agricultural sector is both a victim and a contributor to climate change. It accounts for approximately 21 percent of global greenhouse gas emissions, which equates to about 12 billion metric tons of CO_2 equivalent annually, largely due to livestock production, synthetic fertilizer use, rice cultivation, and deforestation-related land-use changes (U.S. Energy Information Administration [EIA], 2021). Despite efforts to reduce emissions, the pressure to feed an expanding global population, projected to exceed 9.7 billion by 2050, will only increase agricultural intensification and expand cultivated land, placing more strain on ecosystems.

These developments have led to rising levels of food insecurity. The FAO (2022) estimates that over 828 million people were affected by hunger in 2021, a number that has risen due to climate change, conflict, and economic disruption. Without decisive interventions, food production systems may not withstand the growing demand under worsening environmental conditions.

2.2 Emergence of Climate-Smart Agriculture (CSA)

To counter these challenges, the FAO introduced Climate-Smart Agriculture (CSA) in 2010. CSA is not a singular technique but a comprehensive approach that integrates agricultural development and climate responsiveness. It aims to achieve three interconnected objectives: sustainably increasing agricultural productivity and incomes; adapting and building resilience to climate change; and reducing and/or removing greenhouse gas emissions, where possible (FAO, 2024). Figure 2 presents the foundational pillars of CSA, namely productivity, adaptation, and mitigation, highlighting how these goals work together to create sustainable and resilient agricultural systems suited for a changing climate (Bano et al., 2021).

Figure 2. Foundational Pillars of Climate-Smart Agricultural Systems

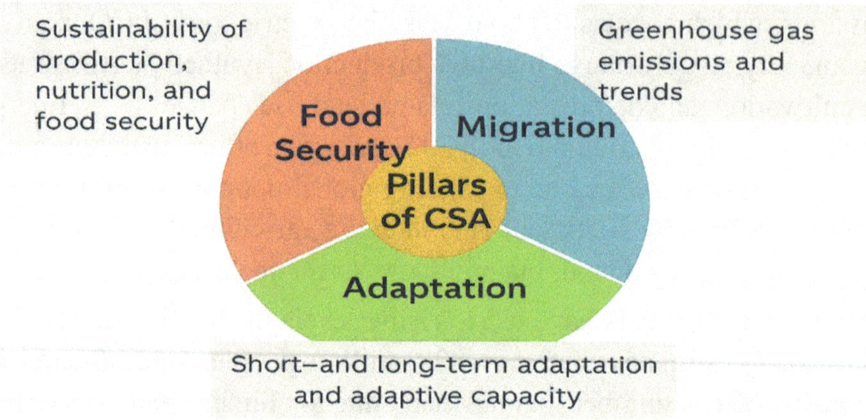

Source: Bano et al., (2021)

CSA includes practices such as conservation tillage, crop rotation, agroforestry, soil fertility management, and the use of drought-tolerant crop varieties. It supports farmers in making context-specific decisions that align with local agro-ecological conditions and socio-economic realities (Vermeulen et al., 2012). CSA is both adaptive and mitigative, meaning it prepares farming systems to endure climate impacts while also contributing to long-term environmental sustainability.

Case studies from Africa, South Asia, and Latin America demonstrate CSA's potential. For example, in Sudan, CSA initiatives like water harvesting, crop diversification, and agroforestry have improved community resilience, food security, and gender equity in rural areas (Elmakki, 2024). Despite these successes, CSA remains underutilized in many parts of the world, largely due to lack of funding, weak policy coordination, limited data, and minimal access to real-time climate information.

2.3 Technological Limitations of Traditional CSA

One of the primary challenges facing CSA is its dependency on local knowledge, manual implementation, and limited access to real-time data. While this approach is often inclusive and sustainable, it can lack the agility and precision needed to deal with rapid-onset climate events such as sudden droughts, pest outbreaks, or rainfall failures. This is particularly evident in regions where early warning systems and weather forecasts are outdated or unavailable.

Additionally, farmers often make decisions based on intuition or tradition, which may not align with changing climate patterns. Without timely access to accurate data, they are unable to adapt in real-time to dynamic agro-ecological conditions (Kogo et al., 2021). The result is crop loss, resource wastage, and economic hardship, particularly for smallholder farmers who lack insurance or financial buffers.

2.4 The Case for Artificial Intelligence in CSA

Artificial Intelligence (AI) offers a solution to many of these limitations. By integrating AI into CSA, farmers, researchers, and policymakers can access high-resolution climate and agricultural data, receive predictive analytics, and implement real-time interventions. AI can process vast amounts of data from multiple sources including satellites, drones, sensors, and historical climate models, and transform it into actionable insights (Vaiyapuri, & Julie, 2025; Assimakopoulos et al., 2025).

AI technologies can also identify trends that are not apparent to the human eye. For example, machine learning algorithms can detect patterns in soil moisture, nutrient content, crop growth stages, or weather anomalies, allowing for more informed decision-making. AI tools can also assist in climate modeling, scenario planning, and risk management, providing farmers with early warnings and adaptive strategies to mitigate losses.

Furthermore, AI makes CSA scalable. Through mobile apps, smart sensors, and AI-driven advisory platforms, even farmers in remote areas can receive personalized guidance, market access, and climate alerts. These tools can bridge the digital divide and make CSA more accessible, equitable, and effective (Sridhar et al., 2023).

2.5 AI for Healthy Food Production

In addition to supporting resilience, AI contributes directly to healthy food production. Through precision agriculture, AI can optimize the application of fertilizers, water, and pesticides, thereby minimizing chemical residues in food and reducing environmental harm. AI systems can monitor nutritional content, forecast market demand for nutrient-rich crops, and enable traceability from farm to fork (Bulut and Arslan, 2022).

Smart agriculture systems supported by AI can also aid in food fortification programs, precision breeding (Bhuiyan et al., 2025), and the cultivation of biofortified crops like iron-rich beans or vitamin A-enhanced sweet potatoes (Antonelli & Donelli, 2023). These interventions play a vital role in combating malnutrition, especially in areas where micronutrient deficiencies are widespread.

Moreover, AI can optimize food supply chains by reducing post-harvest losses, improving cold storage logistics, and enhancing distribution networks. These innovations ensure that more food reaches consumers while preserving quality, safety, and nutritional value.

3. ROLE OF AI IN MODERN AGRICULTURE

3.1 Precision Agriculture and AI Applications

Precision agriculture, also known as site-specific crop management, is a farming method that uses data-driven technologies to optimize agricultural inputs such as water, fertilizers, and pesticides. The core principle is to deliver the right input, at the right time, and in the right amount. AI enhances this approach by analyzing vast and complex data sets to produce actionable insights that drive efficiency, sustainability, and yield improvement (Kumar et al., 2023).

AI-powered platforms can process data from satellite imagery, drones, GPS, soil sensors, and weather forecasts to create real-time maps of field variability. These maps help farmers make informed decisions about which areas require more water or nutrients, where pests are emerging, or when to harvest for maximum yield. For example, convolutional neural networks (CNNs), a type of deep learning algorithm, are widely used in image analysis to detect crop health and diagnose diseases based on visual symptoms (Tugrul, Elfatimi and Eryigit, 2022).

Such AI systems reduce input costs, increase productivity, and minimize environmental degradation. In one case, farms using AI-guided smart irrigation systems reported up to 30 percent water savings without compromising crop yield (Akanbi et al., 2024). Moreover, AI technologies can help monitor soil health over time, allowing farmers to adjust their practices and preserve long-term fertility.

3.2 Automated Systems and Robotics

AI enables the development of autonomous agricultural machinery such as robotic tractors, drone sprayers, and automated harvesters. These machines reduce the need for manual labor, which is especially important in regions facing workforce shortages. Autonomous equipment can work around the clock with high precision and minimal error, improving operational efficiency.

In Japan, AI-powered robots are used to plant rice seedlings with near-perfect spacing, optimizing land use and promoting uniform growth (Li, Mizuki & Imoto, 2025). In the United States, robotic harvesters equipped with machine vision systems are used for picking fruits, recognizing ripe produce, and avoiding damage to plants (Zhu, 2025).

The integration of robotics with AI not only increases productivity but also reduces the occupational health risks associated with manual labor in extreme weather or pesticide-heavy environments. These innovations make agriculture safer, faster, and more reliable.

3.3 AI in Climate Prediction and Risk Management

One of the most powerful applications of AI in agriculture lies in climate risk assessment and prediction. Traditional climate models rely on physical equations and can struggle to incorporate real-time, high-resolution data. AI models, particularly machine learning algorithms, can analyze complex interactions between climate variables, soil characteristics, and crop responses to forecast yield outcomes under different climate scenarios (Assimakopoulos et al., 2025; Kashyap et al., 2024).

AI tools are being used to:

- Predict rainfall patterns and drought probability.
- Model the spread of pests and diseases under climate change.
- Forecast market shocks and food price volatility due to weather events.
- Inform insurance schemes and early warning systems.

For example, IBM's Watson Decision Platform for Agriculture integrates weather data, crop science, and remote sensing to offer real-time recommendations to farmers. In Kenya, AI-based platforms are helping smallholder farmers receive SMS alerts about impending floods or dry spells, allowing them to take preventive measures in advance (Yeung, 2024).

These predictive capabilities are critical for resilience. They give farmers and policymakers a strategic advantage in planning seasonal activities, allocating resources, and preventing crop failures.

3.4 Disease and Pest Detection

Crop losses due to pests and diseases account for nearly 20–40 percent of global agricultural productivity losses annually (FAO, 2025). Traditional detection methods are often reactive, relying on manual inspection or late-stage symptom observation. AI offers a proactive approach by identifying early warning signs through pattern recognition and anomaly detection.

Smartphone-based AI apps allow farmers to photograph plants and receive instant diagnoses along with treatment recommendations. For instance, the PlantVillage Nuru app, developed by Penn State University and CGIAR, uses AI to detect symptoms of crop diseases like cassava mosaic virus and maize lethal necrosis. These tools are especially useful for smallholder farmers who may not have access to agricultural extension services (Mrisho et al., 2020).

AI models trained on regional data can also anticipate pest migration based on climate patterns and vegetation indices. This allows for targeted pesticide application, reducing chemical usage and environmental impact. Moreover, AI can be integrated into drones for aerial scanning of large fields, identifying infected areas and delivering treatments precisely.

3.5 Supply Chain Optimization

AI is revolutionizing agricultural supply chains by improving logistics, reducing waste, and increasing market efficiency. It enables demand forecasting, quality control, and end-to-end traceability. Through machine learning algorithms, platforms can predict demand spikes, recommend optimal distribution routes, and minimize post-harvest losses.

Blockchain technology, when combined with AI, enhances transparency in food systems. It records every transaction, from seed source to shelf, allowing consumers to verify product authenticity and farmers to command better prices for traceable, certified produce (Liu & Li, 2025).

AI-enabled grading systems use computer vision to assess produce quality, ensuring consistency and reducing reliance on subjective human judgment. These systems are widely used in processing facilities to sort fruits, vegetables, and grains based on size, color, and texture.

Smart logistics, backed by AI, can dynamically adjust to disruptions like weather delays, labor shortages, or market fluctuations. This ensures that food reaches markets on time and in good condition, contributing to food security and reduced carbon emissions.

3.6 Farmer Decision Support and Advisory Systems

AI-based platforms serve as virtual advisors, providing farmers with personalized recommendations on everything from seed selection and crop rotation to irrigation and pest control. These platforms often integrate real-time data with historical records to generate location-specific insights.

One such example is the "eSagu" system in India, which uses AI to deliver customized crop management advice via SMS and mobile apps.

Farmers input basic field data and receive step-by-step instructions for maximizing productivity and resilience (Reddy et al., 2007).

These decision-support systems democratize access to knowledge, especially in regions where extension services are limited. They also build farmer confidence, reduce guesswork, and improve adoption of best practices. As literacy and smartphone penetration increase, these tools will become indispensable in modern farming.

4. AI AND CLIMATE RESILIENCE: TECHNOLOGICAL INTEGRATION

4.1 Understanding Climate Resilience in Agriculture

Climate resilience in agriculture refers to the ability of agricultural systems to anticipate, absorb, adapt to, and recover from climatic shocks while maintaining productivity, sustainability, and food security (Douxchamps et al., 2017). Building resilience is particularly crucial for vulnerable farming communities where livelihoods depend heavily on predictable weather conditions. In this context, integrating Artificial Intelligence (AI) into agricultural practices provides transformative opportunities for enhancing resilience.

AI's capacity to process complex datasets, recognize patterns, and predict future conditions allows for dynamic adaptation and proactive response. By analyzing weather patterns, soil health, water availability, and market trends, AI equips farmers, policymakers, and supply chain actors with tools to navigate uncertainty and variability effectively (Aijaz et al., 2025).

4.2 AI-Enabled Weather Forecasting and Early Warning Systems

Traditional weather forecasting relies on static meteorological data and regional models, which often lack accuracy at microclimatic scales. AI, particularly machine learning algorithms, can enhance weather prediction by analyzing large volumes of satellite data, historical climate records, and real-time sensor data. These AI-enhanced forecasts provide localized, short- and long-term predictions with higher accuracy (Eddamiri et al., 2025).

For example, AI-driven early warning systems can alert farmers days or even weeks in advance of extreme events such as heatwaves, floods, droughts, or frost. This allows them to:

- Adjust planting and harvesting schedules.
- Apply protective measures like mulching or irrigation.
- Move livestock or infrastructure to safer areas.
- Secure inputs and insurance coverage ahead of time.

Such systems are already in use in East Africa, where projects like the "Climate Smart Village" initiative deploy AI-powered risk maps and mobile alerts to guide farmer decisions ((Recha et al., 2017)).

4.3 AI for Water Resource Management

Water scarcity is one of the most pressing climate challenges facing agriculture. Efficient water management is critical for improving resilience in drought-prone areas. AI algorithms can analyze real-time soil moisture data, crop water requirements, and evapotranspiration rates to optimize irrigation.

Smart irrigation systems use AI models trained on sensor data and weather forecasts to determine when and how much water to apply. This minimizes water waste, prevents over-irrigation, and reduces the risk of salinization or nutrient leaching (Ali, Tajamul, & Azlan, 2025; Del-Coco et al., 2024).

In regions like Punjab, India, where groundwater depletion poses a serious threat to long-term agricultural sustainability, AI enhanced drip irrigation systems integrating real-time soil moisture sensors and predictive analytics have shown remarkable success. Studies indicate that such systems can reduce water use by 35 to 50% and improve crop yields by 30 to 40%, especially in cereal and horticultural systems (NITI Aayog, 2023). By analyzing environmental variables and automating irrigation schedules, these technologies support data driven decisions, helping farmers manage resources more efficiently and sustainably.

4.4 AI for Soil Health and Nutrient Management

Soil degradation due to over-farming, erosion, and chemical overuse is a major obstacle to resilient agriculture. AI tools can diagnose soil conditions using remote sensing, spectral imaging, and in-situ sensor data. These tools assess pH, nutrient levels, organic content, and compaction to offer tailored soil health recommendations (Bulut and Arslan, 2022).

AI can also guide site-specific fertilizer application, reducing nutrient runoff and greenhouse gas emissions. By avoiding blanket applications, farmers save costs and protect water bodies from contamination.

A notable application is the use of AI to generate dynamic soil fertility maps that change with season and land use. These maps inform both short-term decisions and long-term land restoration strategies.

4.5 Crop Selection and Genetic Resilience through AI

AI contributes to crop improvement by supporting the development of climate-resilient crop varieties. Using genomic datasets, AI algorithms identify genes associated with drought tolerance, pest resistance, and high nutritional value. This speeds up the breeding process and enhances precision in selecting parental lines for hybrid development (Wójcik-Gront, Zieniuk & Pawełkowicz, 2024).

Moreover, AI helps in regional crop suitability modeling. By integrating data on temperature, rainfall, altitude, and soil type, AI systems recommend crops best suited to future climate scenarios. This enables farmers to switch to more resilient crops like millets, legumes, or stress-tolerant maize, which are better adapted to local conditions and climate variability.

Such transitions are critical in regions like the Sahel and the Indo-Gangetic Plain, where climate change is already shrinking the window of opportunity for staple crop cultivation.

4.6 Pest and Disease Surveillance under Changing Climates

As temperatures rise and ecosystems shift, pests and diseases are appearing in new regions. AI models can predict these shifts by correlating environmental changes with pest migration patterns and breeding cycles.

Machine learning models trained on historical and current infestation data allow for early detection and forecasting of outbreaks. Combined with real-time surveillance from drones and satellite imagery, these models form the basis for Integrated Pest Management (IPM) strategies.

AI-supported surveillance reduces reliance on chemical pesticides by promoting timely, targeted interventions. In Kenya, AI models have been instrumental in managing locust invasions by predicting swarm movements and guiding aerial control operations (Gesami & Nunoo, 2024).

4.7 Enhancing Livelihood Resilience and Adaptive Capacity

Beyond technical solutions, AI can support economic and social resilience. AI-powered platforms offer financial services such as climate-indexed insurance, microcredit, and digital wallets, helping farmers cope with losses and invest in adaptive practices. Some digital advisory tools also include gender-sensitive modules, ensuring women farmers, often the most vulnerable to climate impacts, have access to tailored information and support. These systems can track income patterns, recommend livelihood diversification strategies, and connect users to local cooperatives and markets. Through continuous feedback loops and participatory design, AI platforms can evolve with user needs and local realities, thereby enhancing inclusivity and sustainability.

5. CASE STUDIES AND APPLICATION MODELS

5.1 India: AI-Driven Climate Advisory through eSagu and Kisan Call Centres

India, with nearly 70 percent of its rural population reliant on agriculture, faces acute climate variability challenges that threaten food security and rural livelihoods (FAO, 2025). Droughts, erratic monsoons, and heatwaves significantly disrupt food systems. To address this, the Indian government and research institutions have piloted several AI-based advisory platforms that integrate climate-smart agriculture (CSA) principles with real-time AI analytics.

One such example is eSagu, developed by the International Institute of Information Technology (IIIT), Hyderabad. This system combines satellite data, AI algorithms, and field reports to provide localized crop advisories. Farmers receive SMS messages with personalized instructions on irrigation, fertilization, and pest control. Evaluation reports show that eSagu improved yields and reduced chemical inputs, demonstrating significant benefits to participating farmers (Reddy & Ramaraju, 2007).

Another platform, the Kisan Call Centre (KCC), integrates AI-powered voice recognition and machine learning to respond to farmer inquiries in multiple regional languages. AI systems filter the questions, prioritize urgency, and generate advisory responses. These systems not only bridge the digital divide but also provide a scalable model for farmer education and climate adaptation (Rehman et al., 2023).

5.2 Sudan: Climate-Smart Villages and Agro-Ecological Transition

The Climate Smart Village (CSV) model, as promoted by the CGIAR Research Program on Climate Change, Agriculture and Food Security (CCAFS), has been implemented in parts of East Africa including Sudan to integrate traditional knowledge with modern technologies. While not all models directly incorporate AI, some pilot projects utilize digital tools such as decision support systems, mobile apps, and satellite-informed planning to promote drought tolerant crops, water harvesting, and micro irrigation systems adapted to local agroecological conditions (Lim, 2023).

The Sudanese CSVs incorporate smartphone apps that collect real-time data on soil moisture and local weather. These data are fed into AI systems that predict irrigation needs and potential climate risks. The results are shared with local cooperatives via SMS and radio broadcasts.

Climate-Smart Village (CSV) interventions have improved agricultural productivity by 7–18 percent in participating communities and strengthened community cohesion and gender empowerment by involving women in climate monitoring, decision-making, and data collection activities (Challinor et al. 2014).

5.3 Kenya: AI-Powered Locust Control and Pest Surveillance

In East Africa, the 2019–2021 locust infestation posed a severe threat to food security. Traditional pest control methods were insufficient to track the fast-moving swarms. The FAO, in partnership with Penn State University and other agencies, deployed an AI-based system using satellite imagery, weather forecasts, and swarm behavior data to predict locust movement (Yusuf et al., 2024).

This system, combined with drone surveillance, enabled targeted pesticide application and rapid response, limiting crop damage. Moreover, smallholder farmers received AI-generated alerts via the PlantVillage Nuru app, helping them to act preemptively.

This case highlights how AI can support real-time decision-making in pest management, reduce environmental damage, and protect vulnerable farmers during extreme climate events.

5.4 Japan: Robotics and Smart Infrastructure

Japan faces a declining agricultural workforce and increasing labor costs. To maintain food production, the country has embraced AI-enabled robotics in its farming systems. The use of AI-powered autonomous tractors, robotic seeders, and precision harvesters is now common in regions like Hokkaido (Noguchi & Barawid, 2011). As illustrated in Figure 3, the country's projected labor force decline underscores the urgency for automation and smart technologies in agriculture, ensuring productivity in the face of demographic and climate challenges (Kotkin, 2011)

Figure 3. Projected Labor Force Growth

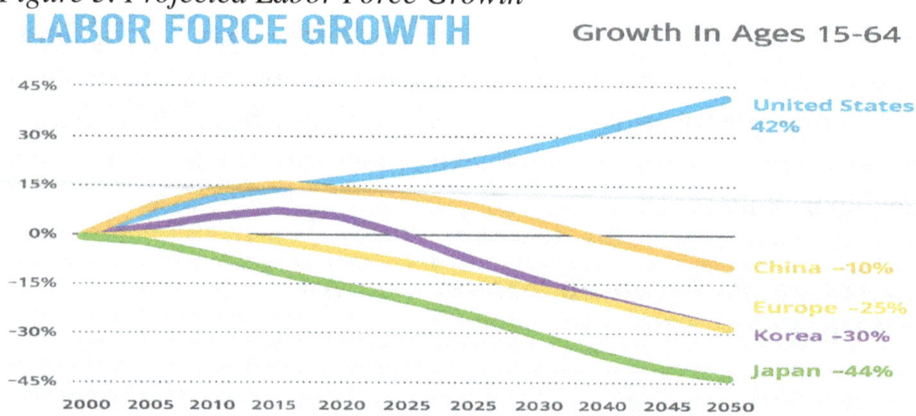

Source: U.S. Census Bureau, International Database; Kotkin, 2011

These robots are guided by GPS and AI vision systems that recognize crop lines and detect obstructions. They operate in diverse weather conditions and can complete tasks with minimal supervision. Additionally, vertical farms equipped with AI control systems manage temperature, lighting, and irrigation, ensuring consistent production of leafy greens in urban settings.

The Japanese case demonstrates how AI not only enhances climate resilience but also compensates for human labor shortages, allowing for year-round production despite aging demographics and adverse weather.

5.5 United States: Climate Risk Management via IBM Watson Decision Platform

In the United States, large-scale farming operations use the IBM Watson Decision Platform for Agriculture, which integrates AI with big data to assist in climate risk management. The platform aggregates information from drones, satellites, and IoT sensors to analyze field variability, predict weather impacts, and optimize planting schedules (Freeman et al., 2019).

For instance, at Paulman Farms in Nebraska, the use of Watson's AI tools helped anticipate rainfall and guide irrigation strategies, leading to water savings and improved productivity during dry seasons. Similarly, in Texas, IBM's "Liquid Prep" system was deployed to support smallholder farmers in drought-prone areas, enabling efficient water use and enhanced crop resilience by providing timely irrigation recommendations.

The U.S. model demonstrates the potential of AI in managing regional climate variability through data-driven insights and responsive strategies.

5.6 China: AI in Supply Chain and Food Traceability

China has actively adopted blockchain and AI technologies to enhance traceability in its agri-food supply chains. In Shandong Province, researchers developed a blockchain-based system integrated with AI-powered monitoring tools to track produce across each stage of the supply chain. This system captures data on temperature, packaging integrity, and logistics from farm to shelf, improving transparency, reducing fraud, and enhancing consumer trust in food safety standards (Mao et al., 2018).

Using computer vision, AI sorts and grades produce for quality control, reducing waste and improving food safety. These systems also support cold chain logistics by dynamically adjusting routes based on traffic and weather conditions, ensuring that perishable food reaches markets without spoilage.

Such applications not only reduce food loss but also build consumer trust, making China's supply chains more resilient to disruptions caused by climate events or pandemics.

5.7 Brazil: Agroecological Innovation with AI

Brazil presents a compelling case for the fusion of artificial intelligence and agroecology, a model that promotes biodiversity, local knowledge, and sustainable practices. Traditionally a global agricultural powerhouse with extensive monoculture farms, Brazil is witnessing a grassroots transition toward regenerative farming systems. AI technologies are being integrated into these models to optimize land use while conserving biodiversity.

Institutes such as Embrapa (Brazilian Agricultural Research Corporation) are deploying AI-powered soil diagnostics and weather forecasting tools tailored to support smallholders in the Cerrado and Amazon transition regions (Oliveira et al., 2021). These systems aid the adoption of agroecological techniques such as cover cropping, no-till farming, and mixed-species cultivation by providing real-time insights based on soil moisture, pest pressure, and microclimatic data. Embrapa's digital platforms and integrated agricultural systems, including their work in the Cerrado-Amazon buffer zones, are critical in guiding sustainable land use, optimizing resource efficiency, and supporting climate-resilient farming practices (Romani et al., 2023).

Moreover, AI assists in evaluating carbon sequestration outcomes of agroforestry systems, enabling local cooperatives to participate in carbon credit programs. This enhances income opportunities while promoting environmental conservation. Through collaborations with Indigenous communities, AI models are trained on local ecological indicators, reinforcing cultural resilience and participatory innovation in sustainable land stewardship.

5.8 Indonesia: Integrating AI in Smallholder Agroforestry

Indonesia, home to one of the world's most complex agroecological zones (Syuaib, 2016), is increasingly turning to AI to support its millions of smallholder farmers who manage agroforestry plots that combine food crops, timber, and biodiversity. These systems are vital for climate mitigation and rural livelihoods but often lack sufficient data and are vulnerable to climate shocks.

AI solutions developed by local startups and research institutes, including Bogor Agricultural University, are using machine learning algorithms and satellite imagery to monitor forest health, crop productivity, and detect illegal deforestation. These tools support national and regional monitoring efforts, achieving up to 95% accuracy in classifying forest cover change and helping identify land-use violations in real time (Irvin et al., 2020;

Farmonaut, 2024). To turn this data into practical help, these systems also send SMS alerts and voice messages in local languages to farmers, offering timely guidance on planting times, pest risks, and expected rainfall.

In places like West Java and Kalimantan, AI-enabled agroforestry platforms and mapping tools are helping improve cocoa and coffee production while keeping tree cover and supporting pollinator habitats. In West Java, coffee farms with diverse shade trees have shown better yields and more pollinators like bees, which help with crop quality and quantity (Campera et al., 2021). In Kalimantan, research by IPB University has used drones and imaging tools to study liberica coffee systems, helping manage land use while preserving biodiversity (IPB University, 2024). These platforms also help farmers work together by collecting their input and supporting group decisions through mobile dashboards.

6. CHALLENGES AND LIMITATIONS

6.1 Data Accessibility and Quality

One of the major limitations of integrating AI in climate-smart agriculture is the availability and quality of agricultural data. For AI models to function effectively, they require access to large volumes of accurate, timely, and region-specific data. In many low- and middle-income countries, such data either do not exist or are fragmented across multiple sources with limited interoperability.

Smallholder farmers often lack digital records on soil quality, crop health, or pest dynamics. Additionally, datasets collected by governments or research institutions may be outdated or inaccessible due to institutional silos. This limits the performance and scalability of AI systems, especially in rural and underdeveloped areas.

Moreover, when datasets are biased or incomplete, AI models may produce inaccurate predictions or recommendations, potentially causing harm rather than benefit. Ensuring data integrity, standardization, and open access is therefore crucial for successful AI deployment in agriculture.

6.2 Digital Divide and Technological Exclusion

Despite the promise of AI, a significant portion of the global farming community remains digitally excluded. Many farmers, particularly in Sub-Saharan Africa and parts of Asia, lack access to smartphones, the internet, or electricity. Women, elderly farmers, and indigenous communities are often disproportionately affected due to socio-cultural, economic, and geographic barriers (Choruma et al., 2024).

Even when the necessary infrastructure exists, low levels of digital literacy hinder adoption. AI-based advisory apps may be too complex or unavailable in local languages, making them inaccessible to the intended users. Without targeted training programs, capacity building, and user-friendly interfaces, the digital divide risks reinforcing existing inequalities in access to information, services, and markets.

6.3 High Costs and Resource Intensity

AI technologies can be expensive to develop, deploy, and maintain. Advanced tools such as satellite imaging, machine vision systems, and AI-powered robots require significant upfront investment, making them out of reach for most small-scale farmers and cooperatives.

Additionally, the infrastructure required to support AI (cloud computing, broadband internet, IoT sensors, and maintenance services) may not be available in remote areas. The cost of hardware, software licenses, and technical support can limit scalability, especially without subsidies or public private partnerships.

High resource intensity can also create a dependency on external providers, making local agricultural systems vulnerable to vendor lock-in and technological obsolescence.

6.4 Ethical and Privacy Concerns

The use of AI in agriculture raises several ethical and privacy-related concerns. As digital tools collect and process personal, geospatial, and economic data, questions emerge about who owns the data and how it is used. Farmers often have limited control or knowledge about how their data is collected, stored, or shared, raising concerns about surveillance, exploitation, and misuse (Kaur et al., 2022).

Moreover, AI models may reflect or reinforce biases present in training data. If algorithms are trained predominantly on data from industrial farms or specific regions, they may produce recommendations that do not apply to diverse agroecological or socio-economic settings. This can lead to ineffective or even harmful outcomes.

Transparent data governance, inclusive algorithm design, and adherence to ethical AI principles are necessary to build trust and safeguard farmer rights.

6.5 Environmental Sustainability and E-Waste

While AI can optimize input use and reduce emissions, the hardware infrastructure that supports AI systems has its own environmental footprint. Data centers consume large amounts of electricity, often sourced from non-renewable energy. The production and disposal of electronic devices also contribute to electronic waste (e-waste), which can be difficult to manage in rural areas.

Furthermore, over-reliance on precision systems may encourage monoculture farming, reducing biodiversity and increasing ecological risks. It is essential to align AI deployment with holistic environmental goals, ensuring that digital solutions contribute to both short-term productivity and long-term ecosystem health.

6.6 Policy and Institutional Gaps

The rapid advancement of AI in agriculture has outpaced the development of policies and regulations to manage it. Many countries lack clear guidelines on digital agriculture, data protection, AI ethics, or intellectual property. This policy vacuum creates uncertainty for developers, investors, and users.

Public institutions may also lack the technical capacity or political will to support AI innovations. Fragmented governance, weak coordination between ministries, and short-term funding cycles hinder the integration of AI into national agricultural strategies. Institutional resistance to change can slow down adoption and reduce effectiveness.

International cooperation, cross-sector partnerships, and capacity building within regulatory bodies are required to create enabling environments for responsible AI deployment in agriculture.

7. POLICY AND GOVERNANCE CONSIDERATIONS

7.1 Need for a Coherent Digital Agriculture Framework

As AI technologies reshape the agricultural sector, there is a pressing need for comprehensive policy frameworks that support their ethical, equitable, and sustainable implementation. Most countries, especially in the Global South, lack dedicated digital agriculture policies. Existing agricultural strategies often do not address digital transformation, and when they do, they tend to focus narrowly on infrastructure without integrating broader socio-economic or environmental considerations.

A coherent framework should align digital innovations with national development goals, climate adaptation strategies, and food security agendas. It should also include provisions for inclusive access, responsible data governance, and capacity building at multiple levels of governance.

Governments need to view AI not just as a technological tool but as part of a systemic shift in how agriculture is practiced, financed, and regulated.

7.2 Data Governance and Farmer Rights

Data sovereignty and farmer rights are central to responsible AI adoption. Farmers often generate the data used by AI systems through their interactions with land, climate, and markets, yet they frequently have limited control over how that data is collected, processed, or shared.

Governments should establish clear regulations to protect:

- Ownership and access rights to farm-level data.

- Consent and transparency in data collection.
- Data anonymization and protection from misuse or commercial exploitation.

Models such as Open Data for Agriculture and Nutrition (GODAN) and the Digital Principles for Development advocate for farmer-centered data governance. These frameworks emphasize participatory design, inclusivity, and open standards, ensuring that digital tools empower rather than disempower farmers.

7.3 Public-Private Partnerships (PPPs)

Governments alone cannot develop and scale AI innovations. Strategic public-private partnerships (PPPs) are critical for resource mobilization, technological development, and service delivery. These partnerships can support:

- Co-investment in infrastructure and digital platforms.
- Collaborative research and development.
- Delivery of farmer advisory services.

Examples include India's collaboration with Microsoft to offer AI-driven soil health cards (Microsoft, 2018). However, PPPs must be governed by accountability mechanisms to prevent monopolies, ensure equitable benefit-sharing, and align commercial interests with public good outcomes.

7.4 Regulatory Standards and Certification

AI systems used in agriculture must meet regulatory standards for performance, reliability, and safety. Governments need to establish certification procedures for AI tools, especially those used for pest management, input optimization, and yield prediction. This ensures:

- Validation of model accuracy.
- Interoperability with existing systems.

- Protection from fraud or manipulation.

Regulatory authorities can collaborate with research institutions, international agencies, and industry actors to define criteria for certification and monitoring. These processes increase user trust and reduce risks associated with untested or low-quality tools.

7.5 Inclusion in National Climate and Development Policies

AI applications should be embedded within national adaptation plans (NAPs), nationally determined contributions (NDCs), and rural development policies. This creates policy coherence and supports funding alignment. By positioning AI as a cross-cutting enabler, countries can unlock international climate finance and donor support.

For instance, Rwanda's NDC includes digital agriculture as a core adaptation strategy, integrating smart farming into broader climate resilience goals. Such integration ensures that AI technologies are not developed in isolation but are part of systemic transformation efforts.

7.6 Capacity Building and Institutional Development

Successful AI deployment depends on institutional readiness. This includes training for policymakers, extension workers, and local government officials to understand and manage AI systems. Technical institutions should offer modules on digital agriculture, data science, and climate modeling.

Investments in universities, vocational centers, and farmer cooperatives can create a pipeline of digital-savvy stakeholders who can support innovation at grassroots and policy levels.

Additionally, multi-stakeholder platforms can facilitate dialogue between developers, users, researchers, and regulators, ensuring that AI solutions are context-appropriate and demand-driven.

8. FUTURE OUTLOOK AND STRATEGIC RECOMMENDATIONS

8.1 Embracing AI as a Pillar of Climate-Resilient Agriculture

As climate stress intensifies and food systems face mounting pressure, AI has emerged as a transformative force capable of reconfiguring agriculture for resilience, sustainability, and equity. Its role transcends yield optimization or input reduction; it is becoming central to planning, forecasting, and managing risks across value chains.

The future of AI in agriculture lies in its integration into national resilience strategies, rural development plans, and sustainable food production policies. Governments, development agencies, and private actors must jointly recognize AI not as a luxury but as an essential infrastructure for ensuring food security in an unpredictable climate.

8.2 Investment in Digital Public Goods

The development of AI-based systems should focus on open-source tools and digital public goods that promote equitable access. Rather than siloed, proprietary platforms, investments must prioritize interoperable technologies that can be adapted across regions and scaled with local content.

For example, AI models for pest prediction or irrigation scheduling should be open-access, enabling researchers and developers to customize them to regional needs. Governments can play a proactive role by funding open datasets, co-developing platforms with universities, and supporting data commons at national and international levels.

This approach ensures that innovation serves the public interest and fosters a culture of collaboration and shared benefit.

8.3 Bridging the Digital Divide

To ensure inclusivity, efforts must focus on bridging the digital divide through:

- Expanding rural connectivity and digital infrastructure.
- Providing affordable smartphones, sensors, and advisory tools.
- Offering digital literacy training, particularly for women, youth, and indigenous communities.
- Supporting intermediaries such as agricultural cooperatives or extension agents to act as digital translators.

Special attention should be given to low-literacy users by developing voice-based AI tools in local languages and integrating culturally relevant interfaces (Patel et al., 2010). Solutions that work offline or in low-bandwidth settings will be particularly valuable in fragile contexts.

8.4 Fostering Multi-Stakeholder Collaboration

The complexity of AI systems requires strong collaboration between governments, farmers, tech developers, researchers, civil society, and donors. Co-creation models that involve farmers in the design and testing of AI tools can increase relevance, uptake, and sustainability.

Platforms like the Global Partnership on Artificial Intelligence (GPAI), CGIAR, and the World Bank's Digital Agriculture program offer mechanisms for knowledge sharing, policy alignment, and joint innovation. These collaborations can facilitate:

- Cross-border research on climate-smart AI.
- Harmonized standards for interoperability.
- Ethical frameworks and safeguards.
- Shared financing mechanisms for infrastructure.

Effective coordination ensures that AI innovations are not fragmented or redundant but part of an integrated agricultural transformation agenda.

8.5 Building Resilient Ecosystems, Not Just Smart Farms

While precision farming and automation are critical, the broader goal should be to build resilient agroecosystems that can sustain livelihoods, biodiversity, and planetary health. AI should support regenerative agriculture, agroecology, and circular systems rather than reinforce extractive models of production.

This involves:

- Designing AI to promote crop diversity, soil regeneration, and climate adaptation.
- Tracking environmental externalities and ecosystem services.
- Aligning AI outputs with sustainability indicators (Kumar et al., 2023).

AI must not only optimize yields but also advance long-term resilience, carbon neutrality, and food sovereignty.

8.6 Enhancing Climate Finance and Risk Sharing Mechanisms

AI can play a pivotal role in enhancing climate finance through risk assessment, monitoring, and verification. For example, AI models can quantify the impact of resilience interventions, enabling results-based payments through climate adaptation funds or carbon credit schemes.

Insurance products can also be enhanced through AI-powered index insurance systems, which monitor rainfall, temperature, or pest incidence to trigger payouts. These innovations protect farmers from loss and increase confidence in climate-smart investments. Researchers argue that integrating machine learning models into government-subsidized insurance structures can reduce uncertainty and severity of climate-induced losses, helping lower premiums and incentivize adoption (Osorio et al., 2024). Similarly, studies adapting AI-driven index models in Colombia found that neural networks performed better than traditional statistical methods for flood-risk payout triggers, improving farmer support mechanisms (Porras-Dimri et al., 2023).

Governments and financial institutions should integrate AI into insurance, subsidies, and incentive programs to reduce climate risk and promote adaptation.

8.7 Ethical AI for Food Justice

Finally, the development of AI for agriculture must be guided by ethical principles that prioritize food justice, equity, and dignity. This includes:

- Preventing algorithmic bias and ensuring fairness.
- Protecting farmer privacy and data sovereignty.
- Ensuring transparency in AI decision-making.
- Avoiding technological determinism or dependency.

AI should serve human-centered development, empowering communities rather than replacing them. Ethical frameworks like the OECD AI Principles and the FAO Ethics of AI for Agriculture should be adopted globally to safeguard rights and promote inclusive innovation.

8.8 Future Trends: Explainable AI, Quantum Tech, and Edge Computing

As AI systems become more advanced, new technologies are being developed that could change how agriculture works. Explainable AI (XAI) is one of these tools. It is designed to make AI decisions easier to understand (Cartolano, Cuzzocrea, & Pilato, 2024). This is especially useful in farming, where AI choices can affect food safety, income, and how land is used. XAI helps farmers understand why a particular recommendation was given, which builds trust and increases the likelihood that they will use the technology.

Quantum computing is another area being explored. It could one day help create better models of climate effects, pest behavior, and plant genetics. While it is still early in development, it may lead to more accurate and faster predictions in the future.

Edge computing refers to processing data directly on a device instead of sending it to the cloud. It gives farmers quick results even in areas with poor internet. Paired with solar-powered tools and sensors, edge computing helps AI work in remote places. This makes it more useful for more farmers.

8.10 Empowering Women and Youth through AI in Agriculture

Women and youth form the backbone of many rural agricultural systems but often face systemic barriers to accessing training, finance, and markets. AI presents a unique opportunity to bridge these gaps through targeted interventions and inclusive platforms.

Several AI-driven mobile apps are now being designed with gender-sensitive interfaces that consider literacy levels, workload, and access to resources. These apps deliver voice-based advisories in local dialects, helping women make decisions about seed selection, pest control, and income diversification.

Youth engagement is also on the rise through agri-tech innovation hubs and digital entrepreneurship programs. In Rwanda, young developers are using open-source AI tools to create crop disease detection apps, soil testing kits, and smart greenhouse systems (Nyakuri et al., 2025). By combining traditional knowledge with AI innovation, youth are rebranding agriculture as a future-ready, tech-enabled sector. These inclusive strategies not only foster innovation but also build resilient agricultural systems that reflect the needs and contributions of all community

9. CONCLUSION

The integration of artificial intelligence into agriculture marks a pivotal turning point in the global effort to build climate-resilient food systems. As the world faces increasing climate variability, rising food insecurity, and unsustainable land-use practices, traditional agricultural methods alone are proving insufficient to meet the scale of these challenges. Artificial intelligence offers a powerful and dynamic solution capable of analyzing vast datasets, forecasting environmental shifts, optimizing resource use, and supporting real-time adaptive strategies. Its role spans across all levels of the agricultural value chain, from field-level monitoring to national policy modeling, making it a transformative tool in the pursuit of sustainable development and food sovereignty. Yet, the real value of artificial intelligence lies not merely in its technical capabilities but in its ethical, inclusive, and equitable deployment across diverse communities and landscapes.

Throughout this chapter, the application of artificial intelligence in agriculture has been examined through both conceptual analysis and practical examples drawn from a wide range of geographies including India, Sudan, Kenya, Japan, the United States, and China. These case studies demonstrate that artificial intelligence, when embedded into contextually appropriate systems and supported by institutional commitment, can increase productivity, reduce input waste, forecast climate risks, and strengthen community resilience. The success stories underscore the need for deliberate strategies that empower smallholder farmers, promote local adaptation, and ensure long-term sustainability. However, these advancements are not without challenges. Persistent barriers such as unequal access to technology, poor data infrastructure, lack of digital literacy, high deployment costs, and the absence of inclusive regulatory frameworks continue to limit the full potential of artificial intelligence in transforming agricultural systems globally.

To truly harness the benefits of artificial intelligence for climate-resilient agriculture, there must be a focused effort on building inclusive digital ecosystems that prioritize marginalized groups and vulnerable regions. Investments in digital infrastructure, development of open-source tools, and integration of artificial intelligence into rural development and climate adaptation strategies are essential steps forward. Public institutions must play a leading role in ensuring that artificial intelligence technologies are not monopolized by private interests but are developed and deployed in ways that reflect collective needs and shared values. This includes strengthening data governance, protecting farmer rights, and embedding ethical principles in every stage of artificial intelligence development and application.

Artificial intelligence should also be viewed as a contributor to long-term ecological sustainability and not merely as a tool for maximizing yields. Its deployment must align with broader goals of biodiversity conservation, soil regeneration, water management, and equitable economic participation. By designing systems that support agroecological models, enable transparent supply chains, and facilitate adaptive learning across farming communities, artificial intelligence can help reshape agriculture into a system that is not only more productive but also more just and resilient. Policymakers, researchers, technology developers, and farmers all have a role to play in realizing this vision, and it is through collective action and shared responsibility that the true potential of artificial intelligence can be unlocked.

In conclusion, artificial intelligence has the capacity to be a cornerstone of climate-smart agriculture and healthy food production. Its transformative promise is not guaranteed by technological sophistication alone but by the human choices that determine its development and use. If guided by principles of equity, sustainability, collaboration, and accountability, artificial intelligence can support a future where agriculture is intelligent, inclusive, and capable of withstanding the complex challenges of a rapidly changing climate.

REFERENCES

Aayog, N. I. T. I. (2023). *Efficiency of Micro-Irrigation in Economizing Water Use in India: Learning from Potential and Under Explored States*. Government of India. https://www.niti.gov.in/sites/default/files/2023-03/Efficiency%20of%20Micro-Irrigation%20in%20economizing%20water%20use%20in%20India.pdf

Aijaz, N., Lan, H., Raza, T., Yaqub, M., Iqbal, R., & Pathan, M. S. (2025). Artificial intelligence in agriculture: Advancing crop productivity and sustainability. *Journal of Agriculture and Food Research*, •••, 101762.

Akanbi, M. B., Banjoko, I. K., Adedotun, K. J., & Raji, A. K. (2024). AI-Powered Smart Irrigation Systems and Solar Energy Integration: A Sustainable Approach to Enhancing Agricultural Productivity in Nigeria. *Journal of Renewable Agricultural Technology Research*.

Ali, A., Tajamul, H., & Azlan, Z. (2025). Smart irrigation technologies and prospects for enhancing water use efficiency for sustainable agriculture. *AgriEngineering*, *7*(4), 106.

Antonelli, M., & Donelli, D. (2023, October). Precision Nutrition and Artificial Intelligence Mobile Apps: A Narrative Review. In *Biology and Life Sciences Forum* (Vol. 29, No. 1, p. 25). MDPI.

Assimakopoulos, F., Vassilakis, C., Margaris, D., Kotis, K., & Spiliotopoulos, D. (2025). AI and related technologies in the fields of smart agriculture: A review. *Information (Basel)*, *16*(2), 100.

Bano, A., Ali, M., Gupta, A., Pathak, N., & Hasan, W. (2021). *Climate Smart Agriculture and Hi-Tech Farming*.

Bhuiyan, M. M. R., Noman, I. R., Aziz, M. M., Rahaman, M. M., Islam, M. R., Manik, M. M. T. G., & Das, K. (2025). Transformation of Plant Breeding Using Data Analytics and Information Technology: Innovations, Applications, and Prospective Directions. *Frontiers in Bioscience-Elite*, *17*(1), 27936.

Bulut, H., & Arslan, S. (2022). Smart Agriculture: IoT and Artificial Intelligence Applications in Modern Farming. *Agricultural Technology Journal*, *34*(2), 101–115.

Campera, M., Budiadi, B., Adinda, E., Ahmad, N., Balestri, M., Hedger, K., & Nekaris, K. A. I. (2021). Fostering a wildlife-friendly program for sustainable coffee farming: The case of small-holder farmers in Indonesia. *Land (Basel)*, *10*(2), 121.

Cartolano, A., Cuzzocrea, A., & Pilato, G. (2024). Analyzing and assessing explainable AI models for smart agriculture environments. *Multimedia Tools and Applications*, *83*(12), 37225–37246.

Challinor, A. J., Watson, J., Lobell, D. B., Howden, S. M., Smith, D. R., & Chhetri, N. (2014). A meta-analysis of crop yield under climate change and adaptation. *Nature Climate Change*, *4*(4), 287–291.

Del-Coco, M., Leo, M., & Carcagnì, P. (2024). Machine learning for smart irrigation in agriculture: How far along are we? *Information (Basel)*, *15*(6), 306.

Douxchamps, S., Debevec, L., Giordano, M., & Barron, J. (2017). Monitoring and evaluation of climate resilience for agricultural development–A review of currently available tools. *World Development Perspectives*, *5*, 10–23.

Eddamiri, S., Bassine, F. Z., Hakam, O., Kambiet, P. L. K., & Chehbouni, A. (2025). Applications of artificial intelligence for forecasting and managing extremes. In *Climate Change and Rainfall Extremes in Africa* (pp. 91-113). Elsevier. Elmakki, O. Climate-Smart Agriculture in Sudan: A Catalyst for Sustainability, Resilience, and Food Security in Humanitarian Efforts.

FAO. (2022). *The State of Food Security and Nutrition in the World 2022*. Food and Agriculture Organization of the United Nations.

Farmonaut. (2024). *Revolutionizing Sustainable Agriculture: How Satellite Imagery and AI Combat Deforestation in Indonesia*. Retrieved from https://farmonaut.com

Food and Agriculture Organization (FAO). (2024). *Climate-Smart Agriculture*. Retrieved July 9, 2025, from https://www.fao.org/climate-smart-agriculture/en/

Food and Agriculture Organization of the United Nations. (2025). Pest and pesticide management: Understanding the context. Retrieved July 9, 2025, from https://www.fao.org/pest-and-pesticide-management/about/understanding-the-context/en/

Freeman, D., Gupta, S., Smith, D. H., Maja, J. M., Robbins, J., Owen, J. S.Jr, & de Castro, A. I. (2019). Watson on the farm: Using cloud-based artificial intelligence to identify early indicators of water stress. *Remote Sensing*, *11*(22), 2645.

Gesami, B. K., & Nunoo, J. (2024). Artificial intelligence in marine ecosystem management: Addressing climate threats to Kenya's blue economy. *Frontiers in Marine Science*, *11*, 1404104.

Irvin, J., Sheng, H., Ramachandran, N., Johnson-Yu, S., Zhou, S., Story, K., . . . Ng, A. Y. (2020). Forestnet: Classifying drivers of deforestation in indonesia using deep learning on satellite imagery. *arXiv preprint arXiv:2011.05479*.

Kashyap, G. S., Kamani, P., Kanojia, M., Wazir, S., Malik, K., Sehgal, V. K., & Dhakar, R. (2024). Revolutionizing agricultureA comprehensive review of artificial intelligence techniques in farming. *Research Advances in Intelligent Computing*, 82-112.

Kaur, J., Hazrati Fard, S. M., Amiri-Zarandi, M., & Dara, R. (2022). Protecting farmers' data privacy and confidentiality: Recommendations and considerations. *Frontiers in Sustainable Food Systems*, *6*, 903230.

Kogo, B. K., Kumar, L., & Koech, R. (2021). Climate change and variability in Kenya: A review of impacts on agriculture and food security. *Environment, Development and Sustainability*, *23*(1), 23–43.

Kotkin, J. (2011, July 11). *America's demographic opportunity*. New Geography. https://www.newgeography.com/content/002492-americas-demographic-opportunity

Kumar, P., Tokas, J., Kumar, N., Lal, M., & Singal, H. R. (2018). Climate change consequences and its impact on agriculture and food security. *International Journal of Chemical Studies*, *6*(6), 124–133.

Li, Q., Mizuki, A., & Imoto, T. (2025). Understanding Japanese rice farmers' preferences for adopting organic agricultural system and uptaking innovative machinery using best-worst scaling method. *Smart Agricultural Technology*, *10*, 100863.

Lim, M. A. (2023). Impact evaluation of climate smart agriculture program investments in food security using machine learning estimators.

Liu, W., & Li, D. (2025). AI-Driven Optimization and Blockchain-Based Traceability for Green Food Supply Chain Safety and Transparency. *Frontiers in Sustainable Food Systems*, *9*, 1597500.

Mao, D., Hao, Z., Wang, F., & Li, H. (2018). Innovative blockchain-based approach for sustainable and credible environment in food trade: A case study in Shandong Province, China. *Sustainability*, *10*(9), 3149.

Microsoft. (2018, September 4). *Digital agriculture: Farmers in India are using AI to increase crop yields*. Microsoft India News. Retrieved July 9, 2025, from https://news.microsoft.com/en-in/features/ai-agriculture-icrisat-upl-india/

Mrisho, L. M., Mbilinyi, N. A., Ndalahwa, M., Ramcharan, A. M., Kehs, A. K., McCloskey, P. C., & Legg, J. P. (2020). Accuracy of a smartphone-based object detection model, PlantVillage Nuru, in identifying the foliar symptoms of the viral diseases of cassava–CMD and CBSD. *Frontiers in Plant Science*, *11*, 590889.

Noguchi, N., & Barawid, O. C., Jr. (2011). Robot farming system using multiple robot tractors in Japan agriculture. *IFAC Proceedings volumes, 44*(1), 633-637.

Nyakuri, J. P., Nkundineza, C., Gatera, O., Nkurikiyeyezu, K., & Mwitende, G. (2025). AI and IoT-powered edge device optimized for crop pest and disease detection. *Scientific Reports*, *15*(1), 22905.

Ofori, S. A., Cobbina, S. J., & Obiri, S. (2021). Climate change, land, water, and food security: Perspectives From Sub-Saharan Africa. *Frontiers in Sustainable Food Systems*, *5*, 680924.

OLIVEIRA, J. D. M., Madari, B. E., Assis, P. C. R., Silveira, A. L. R., Wruck, F. J., JANAÍNA DE MOURA OLIVEIRA, U. F. G., ... & FLAVIO JESUS WRUCK, C. P. A. M. T. (2021). Soil carbon in integrated agricultural production systems in the Cerrado-Amazon transition area.

Osorio, C. P., Leucci, F., & Porrini, D. (2024). Analyzing the relationship between agricultural AI adoption and government-subsidized insurance. *Agriculture*, *14*(10), 1804.

Patel, N., Chittamuru, D., Jain, A., Dave, P., & Parikh, T. S. (2010, April). Avaaj otalo: a field study of an interactive voice forum for small farmers in rural india. In *Proceedings of the SIGCHI conference on human factors in computing systems* (pp. 733-742).

Patil, D. (2024). Artificial Intelligence Innovations In Precision Farming: Enhancing Climate-Resilient Crop Management. *Available at SSRN* 5057424.

Recha, J. W., Kimeli, P., Atakos, V., Radeny, M. A., & Mungai, C. (2017). Stories of success: climate-smart villages in East Africa.

Reddy, P. K., & Ramaraju, G. V. (2007). eSagu18: The Next Generation IT-based Query-less, Cost-effective and Personalized Agro-Advisory System. In *Conference on ICT Solutions for Socio-Economic Development ISED 2007* (p. 80).

Reddy, P. K., Ramaraju, G. V., & Reddy, G. S. (2007, June). eSagu™: a data warehouse enabled personalized agricultural advisory system. In *Proceedings of the 2007 ACM SIGMOD international conference on Management of data* (pp. 910-914).

Reddy, P. P., & Reddy, P. P. (2015). Impacts of climate change on agriculture. *Climate resilient agriculture for ensuring food security*, 43-90.

Rehman, M. Z. U., Raghuvanshi, D., & Kumar, N. (2023). KisanQRS: A deep learning-based automated query-response system for agricultural decision-making. *Computers and Electronics in Agriculture*, *213*, 108180.

Romani, L. A., Evangelista, S. R., Vacari, I., Apolinário, D. R., Vaz, G. J., Speranza, E. A., & Massruhá, S. M. (2023). AgroAPI platform: An initiative to support digital solutions for agribusiness ecosystems. *Smart Agricultural Technology*, *5*, 100247.

Singh, R., & Singh, G. S. (2017). Traditional agriculture: A climate-smart approach for sustainable food production. *Energy, Ecology & Environment*, *2*, 296–316.

Sridhar, A., Ponnuchamy, M., Kumar, P. S., Kapoor, A., Vo, D. V. N., & Rangasamy, G. (2023). Digitalization of the agro-food sector for achieving sustainable development goals: A review. *Sustainable Food Technology*, *1*(6), 783–802.

Sulaiman, R., Chuluunbaatar, D., & Vishnu, S. (2018). Upscaling climate smart agriculture: lessons for extension and advisory services.

Syuaib, M. F. (2016). Sustainable agriculture in Indonesia: Facts and challenges to keep growing in harmony with environment. *Agricultural Engineering International: CIGR Journal*, *18*(2), 170–184.

Tugrul, B., Elfatimi, E., & Eryigit, R. (2022). Convolutional neural networks in detection of plant leaf diseases. *Revista de Agricultura (Piracicaba)*, *12*(8), 1192.

University, I. P. B. (2024). *IPB University Lecturers Conduct Research and Visit Liberika Coffee Commodities in Podo Rukun Village*. https://www.ipb.ac.id/news

U.S. Energy Information Administration. (2021). Annual Energy Outlook 2021. https://www.eia.gov/aeo

Vaiyapuri, T., & Julie, G. (2025). AI and Geospatial Technologies for Climate Change Mitigation: Opportunities, Challenges, and Pathways to Sustainability. *Procedia Computer Science*, *259*, 1346–1355.

Vermeulen, S. J., Campbell, B. M., & Ingram, J. S. (2012). Climate change and food systems. *Annual Review of Environment and Resources*, *37*(1), 195–222.

WILLIAM, B., LIANG, W., & IBRAHIM, A. (2024). Harvesting the Future: Transforming Agriculture with Climate-Smart Innovations.

Wójcik-Gront, E., Zieniuk, B., & Pawełkowicz, M. (2024). Harnessing AI-Powered Genomic Research for Sustainable Crop Improvement. *Agriculture*, *14*(12), 2299.

Yeung, P. (2024, August 21). In Kenya, drought-weary farmers reap rewards of early warnings. Context News. Retrieved July 9, 2025, from https://www.context.news/climate-risks/in-kenya-drought-weary-farmers-reap-rewards-of-early-warnings

Yusuf, I. S., Yusuf, M. O., Panford-Quainoo, K., & Pretorius, A. (2024). A geospatial approach to predicting desert locust breeding grounds in Africa. *arXiv preprint arXiv:2403.06860.*

Zhu, K., Lammers, K., Zhang, K., Arunachalam, C., Bhattacharya, S., Li, J., . . . Li, Z. (2025). Advancement and Field Evaluation of a Dual-arm Apple Harvesting Robot. *arXiv preprint arXiv:2506.05714.*

Chapter 2
The Rise of Smart Farming:
AI's Role in Modern Agriculture

Pooja Dehankar
Ajeenkya D.Y. Patil School of Engineering, India

Kashvi Chaturvedi
https://orcid.org/0009-0005-3192-3400
Ajeenkya D.Y. Patil University, India

Susanta Das
https://orcid.org/0000-0002-9314-3988
Ajeenkya D.Y. Patil University, India

ABSTRACT

AI is becoming more widely acknowledged as a potent instrument for transforming farming methods. A more sustainable & effective food production system is required due to the growing global population, & AI has various benefits in this area. The transformational potential of AI in agriculture is the main topic of this chapter, which also emphasizes how it may improve resource management, address food security, & raise overall agricultural output. A more sustainable & effective agricultural sector is facilitated by AI-powered systems that provide accurate solutions for crop monitoring, disease identification, resource optimization, & increased yield—all of

DOI: 10.4018/979-8-3373-4827-8.ch002

which are crucial given the world's expanding population. Although AI has enormous potential for the agriculture industry, there are many challenges (e.g., need for additional data, creation of accessible & reasonably priced AI solutions), & requirement for farmer education/training. Enhancing AI algorithms, creating fresh uses for AI in agriculture & resolving adoption barriers will be the main goals of future research/development.

1. INTRODUCTION

In the realm of computer science, one key area of study is without a doubt A1. AI is gaining a good deal of popularity because AI problems applicability and potential technological advancements, mainly in regards to challenges that humans or even classical computing paradigms are unable to address (Rich et al., 2009).

According to the United Nations Food & Agriculture Organization a substantial change is required in agriculture due to the massive growth of population by 2050. To solve this unprecedented food scarcity due to climate change and other environmental and human issues, AI will a play a pivotal role. Agricultural production has taken a new turn due to the advancement of AI (Liu, 2020; Rich et al., 2009).

The problem of food shortages and rapid population increase together provide one of the biggest obstacles to global sustainable development. Agricultural systems are under tremendous strain as the world's population grows, particularly in developing nations with scarce resources. These increasing demands are frequently too much for traditional farming practices to handle effectively or sustainably. As a result, new digital technologies like mobile internet, the IoT, and AI are promising ways to change agriculture through smart farming (SF) techniques. In order to make more accurate and well-informed decisions that increase output and decrease resource waste, farmers are using these advancements to gather, transmit, store, and analyze agricultural data in real time. The IoT, which links a variety of sensors to track operational and environmental factors like soil moisture, water levels, irrigation effectiveness, and climate, is at the core of many smart farming systems. The automation of systems like smart irrigation, which combines sensors, controls, and predictive algorithms to guarantee efficient water use and save resources while increasing crop output, depends on this sensor data. Additionally, the deployment of agricultural robots and unmanned aerial vehicles (UAVs) has been made possible by the combination of IoT with AI, ML, DL, and wireless communication technologies. Numerous operations, such as seeding, harvesting, pest control, weed identification, irrigation management, and animal monitoring, can be carried out in real time using these equipments. Smart Decision Support Systems (SDSS) have also grown in importance, especially in developing countries, as they offer real-time analysis of environmental data, crop health, and soil conditions to help improve farm management. By providing farmers with precise recommendations, these systems increase productivity and lessen their negative effects on the environment. However, governments, the commercial sector, and local farming communities must all strongly encourage the effective use of smart farming technologies. To make sure that even smallholder farmers profit from these developments, it is imperative to invest in digital infrastructure, support policies, and training programs. Encouraging smart agriculture is crucial for long-term environmental sustainability, economic resilience in the face of global problems, and improving food security (Mohamed et al., 2021; Rich et al., 2009; Kabato et al., 2025; Titirmare et al., 2024; Rupa, 2025; Mishra & Mishra, 2023; Eli-Chukwu, 2019; Liu, 2020).

Although the agricultural industry has evolved over many years, one of the biggest changes may be the move toward data-driven agriculture and smart farming practices. All areas of agriculture could benefit from the optimization and simplification of agricultural operations through the use of applied AI. Farmers wish to raise their output in terms of quantity and quality, reduce carbon emissions, cut down on labor hours, and boost revenues through data-driven decision assistance and even fully automated tasks. The shift enables modern supply networks and planning models for businesses and government organizations, enhancing the agriculture sector overall. However, there are still several issues that need to be resolved, which could slow down the transfer. Here, businesses, government agencies, and research institutions have essential missions to complete. However, the long-term trend toward more use of AI in agriculture is evident given such powerful incentives. The question now is not whether smart farming will advance further, but rather how the obstacles will be overcome and whose parties stand to gain from its profoundly revolutionary impacts (Jerhamre et al., 2021).

Smart agricultural technology implementation can be expensive and necessitate specific tools, sensors, software, and knowledge (Assimakopoulos et al., 2024). Large-scale data collection and management from several sources can be challenging, necessitating strong infrastructure and data security protocols (Demchenko et al., 2013). To facilitate farmers' decision-making, most agricultural technology advancements enter the farm through equipment. Making better use of natural resources to lower farming trade-offs and satisfy societal demands for sustainable development is the goal of smart farming. While drastically altering human-machine relations, the steadily increasing number of agricultural technologies seeks to help achieve this goal. For farmers and equipment manufacturers alike, this presents both fresh potential and difficulties. As a result, they must increase their knowledge to become proficient with the underutilized smart agricultural instruments. For farmers to use smart farming technologies efficiently, they must have access to education and training (Ritz et al., 2019).

A data tsunami has been caused by the digitization of data in almost every sector of data-driven business. Additionally, digital data handling between machines (M2M) has significantly increased the amount of information. Significant advancements in digital agriculture management applications have influenced information and communication technology (ICT) to benefit farmers and customers alike, and they have also forced technological solutions into rural areas. These issues require greater study, especially in areas like creating user-friendly interfaces, guaranteeing fair access to smart farming technologies, and creating more reliable AI algorithms for agricultural applications (Shaikh et al., 2022).

Sensors and advanced augmented reality visualization techniques are only two examples of the advanced technologies that are currently so frequently employed at such low costs. Consequently, they are already being employed as tools for decision support in a variety of application scenarios. The hardware is ready; the only thing left to do is to incorporate intelligence into it. In this situation, human-centered AI is useful since it involves the domain experts in the decision-making process directly in addition to giving them pertinent and contextual information. Combining artificial intelligence with the natural intelligence of a human expert—who frequently possesses a wealth of experiential knowledge—can be beneficial (Holzinger et al., 2022). As precision agriculture advanced, using technologies like GPS and sensors to better manage resources, the idea of smart farming emerged. With its sophisticated data analysis capabilities, AI integration accelerated the development of smart farming (Karunathilake et al., 2023).

2. PRACTICAL & THEORETICAL IMPLICATIONS

AI systems improve the use of pesticides, fertilizers, and irrigation by analyzing data from sensors (such as weather patterns and soil conditions) and drones. As a result, resources are used more effectively, waste is decreased, and the environmental impact is lessened. Precision farming (or, precision agriculture) increases crop productivity/efficiency using various state-of-the art technologies. Precision agriculture is a cyclic optimization process that requires field data collection, analysis, and evaluation before decisions are made for site-specific field management. This includes the use of sensors, drones, aerial photography, remote sensing technologies, geographic information systems (GIS), global positioning systems (GPS), robotics, and even artificial intelligence (AI) for data collecting and analysis (Yadav & Sidana, 2023; Sharma & Shivandu, 2024).

AI makes it possible to analyze enormous statistics about plant growth, soil health, and climatic trends, leading to a more thorough understanding of intricate agricultural ecosystems (Lu et al., 2024). Farmers may make better judgments by using AI algorithms to create more precise forecast models for insect outbreaks, weather patterns, and crop yields (Delfani et al., 2024).

By evaluating genetic data and finding desired qualities for breeding programs, AI helps speed up the process of agricultural improvement (Rai, 2022). AI can reduce waste and increase output by assisting farmers in more effectively allocating resources (water, fertilizer, and insecticides) (Ashoka et al., 2024).

3. MANAGERIAL IMPLICATIONS

Only when crops are effectively managed can the knowledge they provide be transformed into profitable decisions. Because data is now the most important component of modern agriculture and aids producers in making crucial decisions, Smart Farming is expanding rapidly due to current advancements in data management. To maximize sustainability and production, objective information obtained by sensors yields valuable advantages. The data used in these data-based managed farms can boost productivity by preventing resource waste and environmental contamination. The foundation for future sustainable agriculture is laid by data-driven agriculture, which uses robotic solutions that integrate artificial intelligence approaches. By giving farmers access to real-time data and insights, smart farming technology helps them make better decisions regarding resource management, planting, and harvesting (Saiz-Rubio & Rovira-Más, 2020).

4. PREDICTIVE ANALYSIS AND CROP MANAGEMENT

Predictive analytics and ML increase production forecast accuracy. In order to achieve this, a product-specific machine learning system was developed to evaluate variations across several agricultural data types, including crop health indicators, soil characteristics, and meteorological conditions. With the use of advanced algorithms and data-driven insights, it provides farmers with useful recommendations for improving farming practices to increase agricultural output. The goal is to improve decision-making so that precision agriculture and more sustainable crop management techniques may make use of this data (Singh et al., 2024).

Since agriculture provides food for the nation's people and boosts GDP, it has always been a sector of utmost importance. Several variables, such as soil characteristics, climate, elevation, and irrigation method, affect crop output. In calculating the yield based on this mutual dependence of the components, technological advancements have fallen short. A data-driven model has been created that analyzes and forecasts crops yield over numerous districts' seasons by learning from historical soil and rainfall data. To evolve the expected crop yield, the hybrid neural network model that was designed finds the best combinations of soil factors and combines them with the rainfall pattern in a chosen area. The Time-Series method in Supervised Learning serves as the foundation for the rainfall prediction analysis model. Recurrent neural networks, another subfield of machine learning, are the technology utilized to make the ultimate crop yield estimate. The final projections successfully illustrated the interdependence between soil factors for yield and weather features, thanks to two inter-communicating data-driven models operating at the backend (Kulkarni et al., 2018).

5. CROP HEALTH MONITORING AND DISEASE DETECTION

Farmers have had to cope with pests and diseases from the beginning of agriculture. Although chemical pesticides are a dependable method of managing diseases and pests, their careless application can have detrimental effects on both the environment and human health. Therefore, in order to implement sustainable crop production, it is necessary to identify pest and pathogen infestations at the very beginning of infection so that appropriate management measures can be performed. It can be difficult to identify pests and diseases in standing crops in the outset. Numerous methods, including conventional visual and molecular detection, are employed to invasively identify infections on a very small scale in the field; nevertheless, they necessitate laboratory infrastructure. However, the latest developments in remote sensing and image analysis-based technologies have made it possible to accurately and non-invasively identify the presence of infections and pests over large areas in real time (Abdullah et al., 2023).

The expanding world population, the depletion of natural resources, the loss of arable land, the rapidly shifting climate, and the prevalence of human diseases like Ebola, Lassa, Zika, Nipah, and most recently, the COVID-19 pandemic are some of the issues that the agriculture industry is currently dealing with. The scientific community is under pressure to accomplish Sustainable Development Goal 2 (SDG 2), which aims to end hunger and malnutrition, as a result of these issues, which also threaten global food and nutritional security. For the sake of humanity, technological development is crucial in improving our comprehension of the agricultural system and its interactions at all scales, from the cellular to the green field. For creating precise and accurate datasets to create management tools and models, the application of remote sensing (RS), AI, and ML techniques is very beneficial. These technologies help with soil type identification, water management, fertilizer application optimization, forecasting and early warning model building, crop protection against plant diseases and insect pests, and threat detection, including locusts. Applying RS, AI, and ML algorithms is a potential and revolutionary way to increase agriculture's resistance to biotic and abiotic stresses and attain sustainability to satisfy the demands of the world's expanding population (Aziz et al., 2025).

Farmers are currently interested in Internet of Things (IoT) based wireless sensors that can monitor crop health and weather conditions because of their efficiency and the help that they provide in farmers' work. Apart from this, using artificial intelligence in the field equipment, like autonomous farming equipment, can help them to save the crop from any illness and pests (Qazi et al., 2022).

6. PRECISION IRRIGATION AND RESOURCE MANAGEMENT

One urgent worldwide issue that must be addressed is water scarcity. Only over 31% of the world's population does not experience water stress, according to the UN, indicating that freshwater resources are not managed effectively and are allocated unevenly. Since the human footprint is connected to numerous fields, this phenomenon is both natural and man-made, with the agricultural sector having the biggest influence. According to scientific research, 4.0 technologies are the main forces behind lessening the impact of this industry on priceless resources like soil and water.AI-powered systems can also manage the use of fertilizers and pesticides more precisely, reducing waste and environmental impact. By using a three-layer architecture system to optimize water consumption and reduce soil percolation, the suggested work seeks to improve irrigation management in agriculture by avoiding circumstances where the soil moisture exceeds its capacity point. The assessment of the soil capacity point and, consequently, the establishment of a confidence interval to direct watering decisions are made possible by experimental activities. To generate a consistent dataset for training and testing three distinct machine learning algorithms based on a classification issue to predict the status of the irrigation network, the latter interval, soil and environmental data, and three-day weather forecasts are combined (Preite & Vignali, 2024).

Future developments in energy prices, water shortages, and climate change will spur farmers to create novel agricultural methods. AI must be used immediately to forecast and estimate the ideal water and energy needs for date palm irrigation to accomplish sustainable farming in dry locations. Mohammed et al. (2023) conducted a detailed study of impact of water by changing the setpoints of available water (40, 50, 70, and 80%) on agriculture using various models and data set (extreme gradient boosting (XGBoost), long short-term memory (LSTM) neural network, support vector regression (SVR), and linear regression (LR), Keras library).

For agriculture to become more sustainable and productive, a digital revolution is required. IoT enables real-time crop health and environmental factor (monitoring via sophisticated detectors and helicopters, sending vital data to cloud-based systems. Large datasets and complex data analytics may be handled by these systems because of their strong processing and storage capacities. Prior to the COVID-19 pandemic, the world's food supply was inadequate and insecurely balanced. Short-term profits are given precedence over protecting the planet and its ecosystems in the current agriculture system. Significant structural and automated improvements are needed to build agricultural systems. These challenges could be addressed, though, with the help of intelligent technology and the development of artificial intelligence in agriculture. Artificial intelligence technologies, such as data integration, image processing, and neural networks, have enabled producers to make better decisions by utilizing the most recent data and prediction models. There is hope for an environmentally sustainable future that can feed the world's expanding population thanks to new farming technologies. One area where artificial intelligence is thought to be able to help achieve global sustainability goals is the integration of renewable energy sources. These models' poor learning capabilities make them unsuitable for unpredictable settings. This feedback-integrated technology still performs wonderfully regardless of the season or local conditions. DLISA uses a short-term long-term memory network to predict the amount of soil moisture content for the following day, the time of irrigation, and the amount of water required to irrigate the field (Dhanke et al., 2025).

7. LIVESTOCK MANAGEMENT AND HEALTH MONITORING

The Internet of Things (IoT) has become a game-changing technology in modern agriculture, turning conventional farming methods into more intelligent and effective processes. The monitoring and control of cattle health is one area where IoT applications have shown significant benefits. Because humans rely on animals to meet their needs for meat and dairy products, the health of cattle is extremely important. Modern technologies offer the opportunity to regularly monitor the important health parameters of animals, such as their movement, body condition score, and the amount of food and fluids they consume, rather than responding to diseases after they have spread. By gathering this information and applying modern AI & ML, the farmers may recognize, forecast, and preempt the outbreak of diseases, which can aid them in better decision-making and execute timely interventions. Hence, by embracing IoT technologies, farmers and ranchers can now keep a close check on the well-being of their animals in real time, resulting to greater animal welfare, enhanced production, and increased profitability (Kaur & Virk, 2024).

Assessing livestock wellbeing aids in keeping an eye on the health of the animals to prevent deterioration, detect injuries and stress, and preserve productivity. Because it boosts consumer pressure for a more compassionate transition in animal treatment, it has also become a significant marketing tool. Professionals' and veterinarians' common visual welfare procedures can be expensive and subjective, requiring skilled staff. Recent developments in computer vision, artificial intelligence (AI), and remote sensing have aided in the creation of novel and cutting-edge livestock biometric systems that capture important physiological data linked to animal welfare (Fuentes et al., 2022).

8. SUPPLY CHAIN AND DEMAND FORECASTING

The increasing significance of AI in supply chain management and optimization was emphasized by Kumar et al. (2023). By combining big data and ML approaches, they highlighted how AI plays a critical role in improving and streamlining several supply chain phases, making them more responsive and efficient. AI may assist in finding trends, forecasting demand, cutting down on delays, and enhancing decision-making procedures all throughout the supply chain by evaluating vast amounts of data.

Leong et al. (2023) expanded on this viewpoint by concentrating on the function of AI in the agriculture industry. They noted that by streamlining the production process and the agricultural supply chain, AI has the ability to drastically cut waste and boost productivity. This is accomplished using the Artificial Intelligence of Things (AIoT) concept, which combines AI with IoT technology. AIoT can process and analyze real-time data from a variety of connected devices, including field sensors, equipment, and storage facilities, to improve operational efficiency, cut down on resource waste, and guarantee the timely delivery of agricultural products. Collectively, these studies show how supply chains are becoming smarter and more sustainable due to AI and associated technologies, which are changing both traditional and modern agriculture practices.

9. AUTOMATED WEED CONTROL AND PEST MANAGEMENT

AI-driven weed management is showing promise as a way to assist farmers in increasing yields while lowering floricide use. AI is being utilized in weed management for mapping, predictive modeling, precision weed control, and weed identification. Crops and weeds can be distinguished by training ML algorithms. The following machine learning methods are used to identify weeing: CNN, DCNN, SVM, ANNs, RF classifier, KNN, ShuffleNet-v2, and VGGNet. Advanced laser weeding provides no-till, chemical-free crop weed control. Before they have a chance to harm the crops, laser weeding technology can detect and eliminate weeds at a much earlier stage of their life cycle—before they are even noticeable to the naked eye. It can locate and eradicate weeds in between crops and target them with millimeters of accuracy. Precision, accuracy, cost-effectiveness, environmental preservation, labor efficiency, increased crop yields, time savings, scalability, data-driven insights, climate change resilience, and technical advancement are all benefits of integrating AI into weed management systems (Muchhadiya et al., 2024).

The agricultural sector has an impact on the economy. Farm computerization is the most significant issue and a rapidly expanding topic on a global scale. There is a direct correlation between exponential population growth and increased need for food and opportunities. The farmers' use of traditional farming methods had failed to meet the need. AI has revolutionized the agricultural industry. This technology has protected agricultural production from threats such as population growth, unemployment, warming temperatures, and food insecurity. The information for the 24 classes is collected from a range of agricultural crops to use AI technology installed in drones to identify pests. These technologies limit the use of excessive amounts of water, pesticides, and herbicides, protect soil fertility, improve quality, raise output, and employ labor more efficiently. using drone sensors, irrigation, and farming to apply AI to pest identification. These technologies led to the development of new automation techniques. In addition to providing food, innovative methods gave billions of people new opportunities (Ahuja et al., 2024).

10. REGENERATIVE AGRICULTURE

Regenerative agriculture promotes a comprehensive response to urgent ecological concerns by embracing holistic agriculture and emphasizing biodiversity, soil health, and sustainability. In the modern world, land degradation is a serious issue. Mild to severe erosion affects soils globally. Climate change, biodiversity loss, and soil erosion are major issues that impact the rural sector. By analyzing soil health, weather patterns, and crop performance, AI can provide precise recommendations for planting, irrigation, and fertilization, leading to healthier and more sustainable farming systems. Farmers also benefit from the increasing usage of natural feces and biofertilizers. In addition, farmers should receive technological training so they can use drones and remote sensing to automate their crops. To maintain water levels, cycle supplements, promote fundamental environmental forms, and influence plants, nature, and food security, it is imperative that soil health be prioritized in sustainable gardening. Better nutrient use efficiency, less fertilizer application, and soil health maintenance can be achieved by using nano-fertilizer and adhering to site-specific nutrient management. Thus, regenerative agriculture helps greatly to the battle against land degradation by recovering damaged areas, increasing agricultural production, and promoting long-term soil health while reducing reliance on synthetic fertilizers. Farmers are not aware of this technology, even though it has all the capabilities they need, thus legislative changes and government intervention are needed (Baruah, 2024).

Traditional farming methods might result in decreased output and degraded soil. It is claimed that regenerative agriculture (RA), which emphasizes soil health and carbon sequestration, is the answer to these problems. Maintaining soil cover, minimizing soil disturbance, preserving living roots in the soil throughout the year, enhancing species variety, integrating livestock, and reducing or doing away with the use of synthetic substances (such fertilizers and herbicides) are the core tenets of RA. The overarching goals are to revitalize the land and soil and benefit the larger community on an economic, social, and environmental level. The great majority of growers are hesitant to implement RA despite its alleged advantages since there is insufficient empirical support for its profitability and benefits (Khangura et al., 2023).

11. DIFFUSION OF INNOVATION OR SOCIO-TECHNICAL SYSTEMS THEORY

The Diffusion of Innovation (DOI) theory and the integration of socio-technical systems offer a useful framework for comprehending how farmers might successfully implement and profit from AI tools in agriculture. These frameworks acknowledge that a technology's ability to be successfully implemented depends not just on its technical merits but also on how well it integrates into the larger social, economic, and environmental context. These methods help farmers make better decisions by encouraging them to take into account a range of internal and external aspects, such as socioeconomic situations, local knowledge, resource availability, and climate change. According to the socio-technical systems perspective, technological solutions must be tailored to the requirements and abilities of users, emphasizing the interplay between people, technology, and institutions. This is supported by DOI theory, which explains how new technologies proliferate in communities based on factors such perceived utility, usability, compatibility with current procedures, and the capacity to test and monitor the innovation prior to widespread acceptance. Farmers are more likely to see the benefits of AI technologies and successfully incorporate them into their farming operations when they are presented inside these frameworks. When carefully implemented, AI technology can assist farmers in monitoring crop health, managing climate variability, optimizing inputs like fertilizer and water, and increasing production. Farmers can better address new difficulties in agriculture by coordinating AI deployment with socio-technical and DOI concepts. In the end, this integrated approach encourages farming systems that are more robust, effective, and sustainable (Smit et al., 2024; Sony & Naik, 2020).

12. POLICY AND REGULATORY FRAMEWORKS, AND ACCESS, EQUITY, AND INCLUSIVITY

AI in agriculture has enormous potential to revolutionize farming methods, boost sustainability, and increase production. However, extensive regulations and regulatory frameworks are desperately needed to guarantee that new technologies are used successfully and ethically. These structures must be made to protect farmers' privacy and personal information, stop the unethical or biased use of AI systems, lower the total cost of agricultural production, and guarantee that all farmers, regardless of their location, socioeconomic background, or size of business, have equitable access to digital tools and information. The availability of precise, objective, and context-specific data is crucial for farmers to make well-informed decisions; this necessitates the development of a safe, open, and inclusive information ecosystem.

In this process, government agencies, academic institutions, and non-governmental organizations (NGOs) are all essential. To guarantee that farmers receive timely, accurate, and localized information about soil health, current weather, climate change trends, water availability, crop disease prediction, and the proper application of fertilizers and pesticides to reduce agricultural losses, these stakeholders must cooperate. Access to such information is essential for ensuring sustainability, food security, and climate resilience in agricultural systems in addition to optimizing yields (Tzachor et al., 2022; Lescrauwaet et al., 2022; Maharjan, 2023; Spanaki et al., 2021; Bates et al., 2020; Kaur et al., 2022; Ozor et al., 2025; Fazil et al., 2023; Altieri & Nicholls, 2017).

However, there are a number of obstacles that must be overcome before AI may be successfully applied in agriculture. According to Ali et al. (2024), Wall (2007), Mark (2019), Songol et al. (2021), and Bhat & Huang (2021), a number of obstacles stand in the way of the broad and fair implementation of AI technology. These include the lack of localized datasets to accurately train AI models, the lack of internet connectivity in rural areas, the low level of digital literacy among farmers, and worries about algorithmic bias that could result in unfair recommendations or exacerbate already-existing inequalities. Furthermore, inconsistent deployment and management of AI tools are frequently the result of unclear regulatory requirements and norms. A coordinated and inclusive policy strategy is therefore required to fully achieve the benefits of AI in agriculture. This approach should foster ethical innovation, safeguard farmer rights, close the digital divide, and fortify institutional support for AI-driven agricultural transformation.

13. BIBLIOMETRIC ANALYSIS OF AI IN AGRICULTURE

Fig. 1 shows bibliometric analysis of AI in Agriculture which is created using VOSviewer software and related data has been taken from PubMed. Based on keyword frequency and co-occurrence, the map's several interwoven clusters—each depicted by a different color—indicate thematic intensities.

Figure 1. Bibliometric analysis of AI in agriculture.

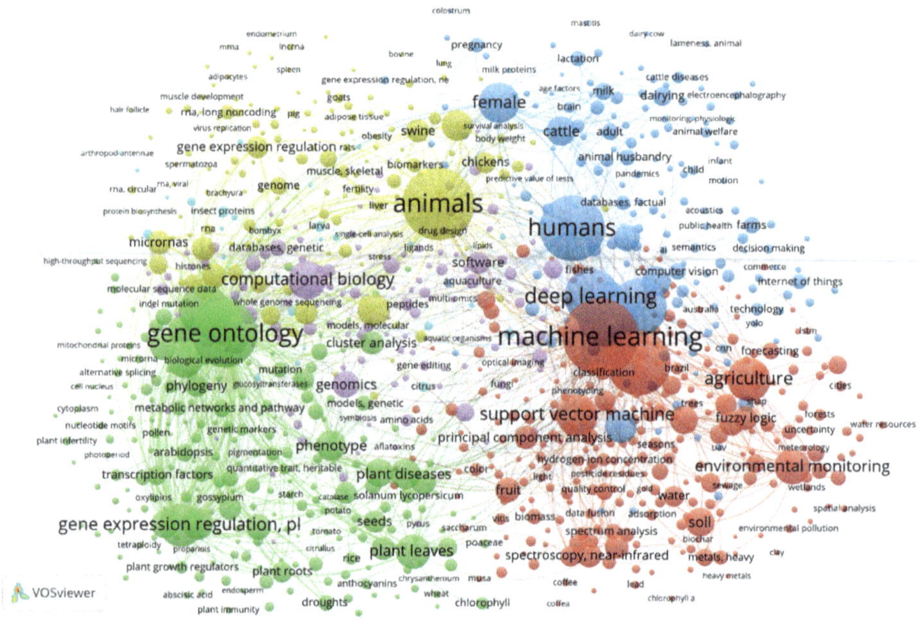

The phrases that are closely related to AI technology and their applications in agriculture and the environment are the focus of the red cluster, which takes up the majority of the bottom-right section of the map. A heavy emphasis on precision agriculture, environmental assessment, and data-driven decision-making processes is indicated by keywords like machine learning, support vector machines, fuzzy logic, spectroscopy, soil, agriculture, environmental monitoring, and remote sensing. The use of advanced sensing and predictive modeling in monitoring soil and crop conditions, controlling irrigation, and resolving sustainability issues is suggested by the close proximity of terms like forecasting, UAV, pesticide detection, water, and infrared.

The top-right quadrant's blue cluster, on the other hand, places more emphasis on research involving humans and animals. Key words like "human," "female," "pregnancy," "lactation," "animal husbandry," and "cattle" imply that AI applications significantly overlap with monitoring animal welfare, livestock management, and reproductive health. Terms like "computer vision," "deep learning," "brain," "public health," and "semantics" imply a wider application of AI in behavioral analysis, health diagnostics, and decision support systems for both humans and animals. Additionally, this cluster displays interactions with real-time monitoring tools and IoT-based systems.

Genomic and biological motifs are the focus of the green cluster, which is noticeable on the left. A strong emphasis on molecular biology, plant genetics, and bioinformatics is shown in keywords such as gene ontology, transcription factors, metabolic networks, microRNAs, phenotype, gene expression regulation, and computational biology. This cluster is strongly focused on plant science, particularly in understanding stress responses, regulatory networks, and genetic mapping utilizing artificial intelligence and computational methods, as seen by the inclusion of terms like plant roots, drought, arabidopsis, genome, and mutation.

The yellow cluster connects animal research with genetics by overlapping between the green and blue areas. The use of bioinformatics and gene expression analysis in veterinary and livestock research is demonstrated by keywords such as animals, swine, chickens, spleen, adipose tissue, muscle development, and biomarkers. This cluster's cross-disciplinary nature is further reinforced by the existence of databases and computational biology, which connect molecular data to organism-level characteristics and production results.

In the center is a smaller purple cluster that contains keywords from AI modeling, phenotypic prediction, and genomics. A computational center where data science technologies are being used to examine biological datasets that link the plant and animal domains is indicated by terms like genomics, cluster analysis, biomarkers, and software.

Machine learning is a key node in the map, with close ties to almost all theme areas, illustrating its integrative function in environmental, agricultural, and biological research. Additional bridging phrases that indicate interdisciplinary similarities include deep learning, genomics, phenotype, and environmental monitoring. These concepts serve as connections between different clusters. The intricate and dynamic character of AI research in the agricultural and biological sciences, which is marked by high interconnection, topic diversity, and methodological integration, is demonstrated by the dense network of links and different node sizes.

14. CASE STUDIES

In order to forecast crop yields and determine which crops are best to sow in particular seasons and regions, AI models can examine previous data. The best crop production forecast model for the Midwestern United States (US) is developed by comparing various AI algorithms. The July-August (JA) database proved to be the most effective in forecasting maize and soybean yields after the authors tested three distinct time periods to investigate the effects of phenology. The authors looked at six distinct AI algorithms for predicting agricultural yield. Next, a thorough and objective comparison of the AI models is carried out. The authors specifically refined the deep neural network (DNN) model to guarantee optimal settings for the layer structure, cost function, optimizer, activation function, and drop-out ratio (Kim et al., 2019).

Due to its many effects on the nation's economy, food security, rural development, and export potential, agriculture is given a lot of weight in India. Agriculture, which provides food security for the country's large population and is the main source of income for millions of people, employs a sizable share of the labor force. AI has the potential to transform agriculture and increase production, sustainability, and efficiency in a number of ways. Among the uses of AI in agriculture that are studied include insect control, irrigation systems, crop monitoring, and predictive analytics (Patil et al., 2023).

For many Africans living in poverty, small farms are vital to their existence. In an attempt to fight poverty in the region, there has been a lot of interest in assisting these farmers in increasing their productivity and making a profit. Examples of such interventions include educating farmers and exposing them to new agricultural practices and goods, like fertilizers or new crop kinds. It is uncertain how these interventions will impact food security and economic outcomes, even if governments and international donors have made large financial expenditures in these programs. More top-notch research is needed to assess different training, innovation, and new technology solutions. Further research is needed to ascertain whether such therapies have long-term, sustainable impacts and whether they could harm farmers or their communities (Stewart et al., 2015).

In India, crop monitoring is being revolutionized by drones fitted with hyperspectral and multispectral sensors (Bahuguna et al., 2021). The vast amounts of imagery that these flying sensors gather are analyzed by artificial intelligence algorithms to find early warning indicators of disease outbreaks, pest infestations, nutritional deficits, and water stress (Sharada et al., 2025).

India employs AI to address climate change and resource scarcity in agriculture (Mor et al., 2021). IoT-enabled technologies, which enhance farmer livelihoods, productivity, and sustainability, are one example of the nation's commitment to AI-driven agricultural revolution (Ahamed, 2024).

15. CHALLENGES

Exorbitant costs, poor infrastructure, resistance to change, data limitations, and ethical quandaries are some of the main barriers preventing the mainstream application of AI in agriculture (Ali et al., 2024). These obstacles prevent widespread adoption, especially in emerging nations and among small-scale farmers (Wall, 2007). To solve these problems, a multipronged approach is required, which includes developing more affordable and easily available AI solutions, improving data infrastructure and quality, providing farmers with support and training, and establishing ethical guidelines for AI use (Mark, 2019). Some farmers are hesitant to adopt new technologies because of their familiarity with traditional ways and their ignorance about artificial intelligence.

This resistance may prevent the adoption of AI technologies, even when they offer potential benefits (Songol et al., 2021). Serious concerns regarding data security and privacy are raised by the collection and analysis of sensitive agricultural data by AI systems. Farmers need to know that their data will be handled responsibly and safely. The potential environmental impacts of AI-driven agriculture also raise ethical questions (Bhat & Huang, 2021).

16. FUTURE SCOPE

AI will enable farmers to make well-informed decisions by analyzing real-time data on soil conditions, weather trends, and crop health (Mishra & Mishra, 2024). AI may maximize the use of resources, such as pesticides, fertilizers, and water, resulting in less waste and a less environmental effect (Onyeaka et al., 2023). Broad-spectrum insecticides are not as necessary thanks to AI-powered systems that can monitor crop health, identify illnesses and pests early, and suggest efficient treatments (Upadhyay et al., 2025). Eissa (2024) mentioned that human labor can be replaced by AI tools (robotics), this saving human resources. According to Jha et al. (2019) AI can be utilized by farmers for planning, planting and harvesting efficiently the crops. AI can be utilized to monitor the health of the crops, thus reducing the losses of crops (Amulothu et al., 2024). AI can also help the farmers to adopt the change in the climate and plan accordingly for optimum production of crops, and therefore can support sustainable farming with optimum yield (Gryshova et al., 2024; Adewusi et al., 2024; Malhotra & Firdaus, 2022).

17. CONCLUSION

AI plays a very important role in agriculture sector. The absence of efficient irrigation systems, weeds, difficulties monitoring plants because of crop height, and harsh weather are some of the problems facing the agriculture sector. However, technology can improve performance, which will help to solve these issues. Various AI-driven methods, like GPS-assisted automatic irrigation and remote sensors for detecting soil moisture content, can enhance it. Farmers had to overcome the significant number of crops lost during the weeding process by using precision weeding techniques.

These self-governing robots not only increase productivity but also lessen the need for needless herbicides and insecticides.

In addition, farmers may use drones to efficiently spray pesticides and herbicides on their farms, and plant monitoring is no longer a hassle. First, man-made intelligence can be used to understand agribusiness problems, such as limitations of jobs and resources. In traditional methods, obtaining agricultural attributes such as plant height, soil texture, and content required a great deal of work; manual testing consequently took place, which was tiresome. With the help of the many technologies that were looked at, rapid and non-destructive high throughput phenotyping would take place, with the benefits of favorable and flexible activity, on-demand information access, and spatial objectives.

AI has the potential to completely transform agriculture and provide a more productive, efficient, and sustainable food system. Farmers can overcome obstacles, increase crop yields, and support global food security by utilizing AI.

REFERENCES

Abdullah, H. M., Mohana, N. T., Khan, B. M., Ahmed, S. M., Hossain, M., Islam, K. S., Redoy, M. H., Ferdush, J., Bhuiyan, M., Hossain, M. M., & Ahamed, T. (2023). Present and future scopes and challenges of plant pest and disease (P&D) monitoring: Remote sensing, image processing, and artificial intelligence perspectives. *Remote Sensing Applications: Society and Environment*, *32*, 100996. DOI: 10.1016/j.rsase.2023.100996

Adewusi, A. O., Asuzu, O. F., Olorunsogo, T., Iwuanyanwu, C., Adaga, E., & Daraojimba, D. O. (2024). AI in precision agriculture: A review of technologies for sustainable farming practices. *World Journal of Advanced Research and Reviews*, *21*(1), 2276–2285.

Ahamed, T. (2024). *IoT and AI in Agriculture: Smart Automation Systems for Increasing Agricultural Productivity to Achieve SDGs and Society 5.0*. Springer Nature.

Ahuja, S., Manjunath, H. R., Alam, I., & Rastogi, A. (2024). The Role of AI in Modern Farming: Precise Pest Management and Optimal Water Use. *International Journal of Chemical and Biochemical Science*, *25*(13), 249–255.

Ali, G., Mijwil, M. M., Buruga, B. A., Abotaleb, M., & Adamopoulos, I. (2024). A survey on artificial intelligence in cybersecurity for smart agriculture: State-of-the-art, cyber threats, artificial intelligence applications, and ethical concerns. *Mesopotamian Journal of Computer Science*, *2024*, 53–103.

Altieri, M. A., & Nicholls, C. I. (2017). The adaptation and mitigation potential of traditional agriculture in a changing climate. *Climatic Change*, *140*, 33–45.

Amulothu, D. V. R. T., Rodge, R. R., Hasan, W., & Gupta, S. (2024). Machine Learning for Pest and Disease Detection in Crops. In *Agriculture 4.0* (pp. 111–132). CRC Press.

Ashoka, P., Devi, B. R., Sharma, N., Behera, M., Gautam, A., Jha, A., & Sinha, G. (2024). Artificial Intelligence in Water Management for Sustainable Farming: A Review. *Journal of Scientific Research and Reports, 30*(6), 511–525.

Assimakopoulos, F., Vassilakis, C., Margaris, D., Kotis, K., & Spiliotopoulos, D. (2024). The implementation of "smart" technologies in the agricultural sector: A review. *Information (Basel), 15*(8), 466.

Aziz, D., Rafiq, S., Saini, P., Ahad, I., Gonal, B., Rehman, S. A., Rashid, S., Saini, P., Rohela, G. K., Aalum, K., & Nabila Iliya, M. (2025). Remote sensing and artificial intelligence: Revolutionizing pest management in agriculture. *Frontiers in Sustainable Food Systems, 9*, 1551460.

Baruah, R. (2024). Restoring Soil Health of Degraded Lands Through Eco-Friendly Nutrient Management Practices in Regenerative Agriculture. In *Regenerative Agriculture for Sustainable Food Systems* (pp. 237–269). Springer Nature Singapore.

Bates, D. W., Auerbach, A., Schulam, P., Wright, A., & Saria, S. (2020). Reporting and implementing interventions involving machine learning and artificial intelligence. *Annals of Internal Medicine, 172*(11, Supplement), S137–S144. PMID: 32479180

Bhat, S. A., & Huang, N. F. (2021). Big data and ai revolution in precision agriculture: Survey and challenges. *IEEE Access : Practical Innovations, Open Solutions, 9*, 110209–110222.

Delfani, P., Thuraga, V., Banerjee, B., & Chawade, A. (2024). Integrative approaches in modern agriculture: IoT, ML and AI for disease forecasting amidst climate change. *Precision Agriculture, 25*(5), 2589–2613.

Demchenko, Y., Ngo, C., de Laat, C., Membrey, P., & Gordijenko, D. (2013). Big security for big data: Addressing security challenges for the big data infrastructure. In *Workshop on secure data management* (pp. 76-94). Cham: Springer International Publishing.

Dhanke, J. A., Srivastava, D., Menaga, D., Raj, R., Kumar, K. V., Jangir, P., & Mani, P. (2025). Climate-Based AI-Powered Precision Irrigation: Sustainably Smart Agriculture Frameworks for Maximum Crop Yields. *Remote Sensing in Earth Systems Sciences*, *8*(1), 161–172.

Eli-Chukwu, N. C. (2019). Applications of artificial intelligence in agriculture: A review. *Engineering, Technology &. Applied Scientific Research*, *9*(4).

Fazil, A. W., Hakimi, M., & Shahidzay, A. K. (2023). A comprehensive review of bias in AI algorithms. *Nusantara Hasana Journal*, *3*(8), 1–11.

Fuentes, S., Viejo, C. G., Tongson, E., & Dunshea, F. R. (2022). The livestock farming digital transformation: Implementation of new and emerging technologies using artificial intelligence. *Animal Health Research Reviews*, *23*(1), 59–71. PMID: 35676797

Gryshova, I., Balian, A., Antonik, I., Miniailo, V., Nehodenko, V., & Nyzhnychenko, Y. (2024). Artificial intelligence in climate smart in agricultural: Toward a sustainable farming future. *Access J*, *5*(1), 125–140.

Holzinger, A., Saranti, A., Angerschmid, A., Retzlaff, C. O., Gronauer, A., Pejakovic, V., Medel-Jimenez, F., Krexner, T., Gollob, C., & Stampfer, K. (2022). Digital transformation in smart farm and forest operations needs human-centered AI: Challenges and future directions. *Sensors (Basel)*, *22*(8), 3043. PMID: 35459028

Jerhamre, E., Carlberg, C. J. C., & Van Zoest, V. (2021). Exploring the susceptibility of smart farming: Identified opportunities and challenges. *Smart Agricultural Technology*, *2*, 100026. DOI: 10.1016/j.atech.2021.100026

Jha, K., Doshi, A., Patel, P., & Shah, M. (2019). A comprehensive review on automation in agriculture using artificial intelligence. *Artificial Intelligence in Agriculture*, *2*, 1–12.

Kabato, W., Getnet, G. T., Sinore, T., Nemeth, A., & Molnár, Z. (2025). Towards climate-smart agriculture: Strategies for sustainable agricultural production, food security, and greenhouse gas reduction. *Agronomy (Basel)*, *15*(3), 565.

Karunathilake, E. M. B. M., Le, A. T., Heo, S., Chung, Y. S., & Mansoor, S. (2023). The path to smart farming: Innovations and opportunities in precision agriculture. *Agriculture*, *13*(8), 1593.

Kaur, D., & Virk, A. K. (2024). Leveraging IoT for precision health monitoring in livestock with artificial intelligence. In *Data-driven farming* (pp. 1–18). Auerbach Publications.

Kaur, J., Hazrati Fard, S. M., Amiri-Zarandi, M., & Dara, R. (2022). Protecting farmers' data privacy and confidentiality: Recommendations and considerations. *Frontiers in Sustainable Food Systems*, *6*, 903230.

Khangura, R., Ferris, D., Wagg, C., & Bowyer, J. (2023). Regenerative agriculture—A literature review on the practices and mechanisms used to improve soil health. *Sustainability*, *15*(3), 2338.

Kim, N., Ha, K. J., Park, N. W., Cho, J., Hong, S., & Lee, Y. W. (2019). A comparison between major artificial intelligence models for crop yield prediction: Case study of the midwestern United States, 2006–2015. *ISPRS International Journal of Geo-Information*, *8*(5), 240.

Kulkarni, S., Mandal, S. N., Sharma, G. S., & Mundada, M. R. (2018). Predictive analysis to improve crop yield using a neural network model. In *2018 international conference on advances in computing, communications and informatics (ICACCI)* (pp. 74-79). IEEE.

Leong, Y. M., Lim, E. H., Subri, N. F. B., & Jalil, N. B. A. (2023). Transforming agriculture: Navigating the challenges and embracing the opportunities of artificial intelligence of things. In *2023 IEEE International Conference on Agrosystem Engineering, Technology & Applications (AGRETA)* (pp. 142-147). IEEE.

Lescrauwaet, L., Wagner, H., Yoon, C., & Shukla, S. (2022). Adaptive legal frameworks and economic dynamics in emerging technologies: Navigating the intersection for responsible innovation. *Law and Economics*, *16*(3), 202–220.

Liu, S. Y. (2020). Artificial intelligence (AI) in agriculture. *IT Professional*, *22*(3), 14–15.

Lu, K., Yan, W., & Tan, H. (2024). Big data-based soil health and sustainable agriculture: Analysis of structure, nutrient, and microbial interactions. *Geographical Research Bulletin, 3*, 263–281.

Maharjan, P. (2023). The Role of Artificial Intelligence-Driven Big Data Analytics in Strengthening Cybersecurity Frameworks for Critical Infrastructure. *Global Research Perspectives on Cybersecurity Governance, Policy, and Management, 7*(11), 12–25.

Malhotra, K., & Firdaus, M. (2022). Application of artificial intelligence in IoT security for crop yield prediction. *ResearchBerg Review of Science and Technology, 2*(1), 136–157.

Mark, R. (2019). Ethics of using AI and big data in agriculture: The case of a large agriculture multinational. *The ORBIT Journal, 2*(2), 1–27.

Mishra, H., & Mishra, D. (2023). Artificial intelligence and machine learning in agriculture: Transforming farming systems. *Res. Trends Agric. Sci, 1*, 1–16.

Mishra, H., & Mishra, D. (2024). AI for Data-Driven Decision-Making in Smart Agriculture: From Field to Farm Management. In *Artificial Intelligence Techniques in Smart Agriculture* (pp. 173–193). Springer Nature Singapore.

Mohamed, E. S., Belal, A. A., Abd-Elmabod, S. K., El-Shirbeny, M. A., Gad, A., & Zahran, M. B. (2021). Smart farming for improving agricultural management. *The Egyptian Journal of Remote Sensing and Space Sciences, 24*(3), 971–981.

Mor, S., Madan, S., & Prasad, K. D. (2021). Artificial intelligence and carbon footprints: Roadmap for Indian agriculture. *Strategic Change, 30*(3), 269–280.

Muchhadiya, R. M., Kumawat, P. D., Gohil, B. S., & Sakarvadia, H. L. (2024). Precision weed management using artificial intelligence tools and techniques for sustainable agriculture. *Agriculture & Food e-Newsletter, 6*(1).

Onyeaka, H., Tamasiga, P., Nwauzoma, U. M., Miri, T., Juliet, U. C., Nwaiwu, O., & Akinsemolu, A. A. (2023). Using artificial intelligence to tackle food waste and enhance the circular economy: Maximising resource efficiency and minimising environmental impact: A review. *Sustainability*, *15*(13), 10482.

Ozor, N., Nwakaire, J., Nyambane, A., Muhatiah, W., & Nwobodo, C. (2025). Enhancing Africa's agriculture and food systems through responsible and gender inclusive AI innovation: Insights from AI4AFS network. *Frontiers in Artificial Intelligence*, *7*, 1472236. PMID: 39917548

Patil, S., Pisal, D., Mirji, H., & Phalke, V. (2023). AI-Powered Innovations in Agriculture: Transforming Farming Practices and Yield Optimization. *The Indian Journal of Technical Education*, *46*, 126–134.

Preite, L., & Vignali, G. (2024). Artificial intelligence to optimize water consumption in agriculture: A predictive algorithm-based irrigation management system. *Computers and Electronics in Agriculture*, *223*, 109126.

Qazi, S., Khawaja, B. A., & Farooq, Q. U. (2022). IoT-equipped and AI-enabled next generation smart agriculture: A critical review, current challenges and future trends. *IEEE Access : Practical Innovations, Open Solutions*, *10*, 21219–21235.

Rai, K. K. (2022). Integrating speed breeding with artificial intelligence for developing climate-smart crops. *Molecular Biology Reports*, *49*(12), 11385–11402. PMID: 35941420

Rich, E., Knight, K., & Nair, S. (2009). *Artificial Intelligence*. McGraw Hill Company.

Ritz, S., Rizzo, D., Fourati-Jamoussi, F., Dantan, J., Combaud, A., & Dubois, M. J. (2019). *Training in agricultural technologies: a new prerequisite for smart farming. In 3rd Rendez-Vous Techniques AXEMA.*

Rupa, G. (2025). AI-Powered Solutions for Crop Disease Prediction, Market Stability, and Sustainable Fertilizer Use to Enhance Farmer Productivity. [IJARMT]. *International Journal of Advanced Research and Multidisciplinary Trends*, *2*(2), 248–259.

Saiz-Rubio, V., & Rovira-Más, F. (2020). From smart farming towards agriculture 5.0: A review on crop data management. *Agronomy (Basel), 10*(2), 207.

Shaikh, T. A., Rasool, T., & Lone, F. R. (2022). Towards leveraging the role of machine learning and artificial intelligence in precision agriculture and smart farming. *Computers and Electronics in Agriculture, 198*, 107119.

Sharada, K., Choudhary, S. L., Harikrishna, T., Dixit, R. S., Suman, S. K., Ayyappa Chakravarthi, M., & Bhagyalakshmi, L. (2025). GeoAgriGuard: AI-Driven Pest and Disease Management with Remote Sensing for Global Food Security. *Remote Sensing in Earth Systems Sciences*, •••, 1–14.

Sharma, K., & Shivandu, S. K. (2024). Integrating artificial intelligence and internet of things (IoT) for enhanced crop monitoring and management in precision agriculture. *Sensors International*, •••, 100292.

Singh, I., Sharma, M., & Tiwari, Y. (2024). Predictive Analytics for Crop Management: Leveraging Machine Learning for Yield Analysis. In *2024 Second International Conference on Advanced Computing & Communication Technologies (ICACCTech)* (pp. 558-563). IEEE.

Smit, D., Eybers, S., van der Merwe, A., Wies, R., Human, N., & Pielmeier, J. (2024). Integrating the TOE framework and DOI theory to dissect and understand the key elements of AI adoption in sociotechnical systems. *South African Computer Journal = Suid-Afrikaanse Rekenaartydskrif, 36*(2), 159–176.

Songol, M., Awuor, F., & Maake, B. (2021). Adoption of artificial intelligence in agriculture in the developing nations: A review. *Journal of Language. Technology & Entrepreneurship in Africa, 12*(2), 208–229.

Sony, M., & Naik, S. (2020). Industry 4.0 integration with socio-technical systems theory: A systematic review and proposed theoretical model. *Technology in Society, 61*, 101248.

Spanaki, K., Karafili, E., & Despoudi, S. (2021). AI applications of data sharing in agriculture 4.0: A framework for role-based data access control. *International Journal of Information Management, 59*, 102350.

Stewart, R., Langer, L., Da Silva, N. R., Muchiri, E., Zaranyika, H., Erasmus, Y., Randall, N., Rafferty, S., Korth, M., Madinga, N., & De Wet, T. (2015). The effects of training, innovation and new technology on African smallholder farmers' economic outcomes and food security: A systematic review. *Campbell Systematic Reviews*, *11*(1), 1–224.

Titirmare, S., Margal, P. B., Gupta, S., & Kumar, D. (2024). AI-powered predictive analytics for crop yield optimization. In *Agriculture 4.0* (pp. 89-110). CRC Press. This leads to more efficient resource use, reduced waste, and improved sustainability.

Tzachor, A., Devare, M., King, B., & Avin, S., & Ó hÉigeartaigh, S. (2022). Responsible artificial intelligence in agriculture requires systemic understanding of risks and externalities. *Nature Machine Intelligence*, *4*(2), 104–109.

Upadhyay, A., Patel, A., Patel, A., Chandel, N. S., Chakraborty, S. K., & Bhalekar, D. G. (2025). Leveraging AI and ML in Precision Farming for Pest and Disease Management: Benefits, Challenges, and Future Prospects. *Ecologically Mediated Development: Promoting Biodiversity Conservation and Food Security*, 511-528.

VOSviewer Software. *Retrieved, 2025.* https://www.vosviewer.com/

Wall, P. C. (2007). Tailoring conservation agriculture to the needs of small farmers in developing countries: An analysis of issues. *Journal of Crop Improvement*, *19*(1-2), 137–155.

Yadav, N., & Sidana, N. (2023). Precision Agriculture Technologies: Analysing the Use of Advanced Technologies, Such as Drones, Sensors, and GPS, In Precision Agriculture for Optimizing Resource Management, Crop Monitoring, and Yield Prediction. *Journal of Advanced Zoology, 44.*

ADDITIONAL READING

Ahmad, A., Liew, A. X. W., Venturini, F., Kalogeras, A., Candiani, A., Di Benedetto, G., Ajibola, S., Cartujo, P., Romero, P., Lykoudi, A., De Grandis, M. M., Xouris, C., Lo Bianco, R., Doddy, I., Elegbede, I., Labate, G. F. D., Del Moral, L. F. G., & Martos, V. (2024). AI can empower agriculture for global food security: Challenges and prospects in developing nations. *Frontiers in Artificial Intelligence, 7*. Advance online publication. DOI: 10.3389/frai.2024.1328530 PMID: 38726306

Bahuguna, S., Anchal, S., Guleria, D., & Devi, M., Meenakshi, Kumar, D., Kumar, R., Murthy, P.V.S. & Kumar, A. (2021). Unmanned aerial vehicle-based multispectral remote sensing for commercially important aromatic crops in India for its efficient monitoring and management. *Photonirvachak (Dehra Dun)*, •••, 1–11.

Balafoutis, A. T., Beck, B., Fountas, S., Tsiropoulos, Z., Vangeyte, J., van der Wal, T., Soto-Embodas, I., Gómez-Barbero, M. & Pedersen, S. M. (2017). Smart farming technologies–description, taxonomy and economic impact. *Precision agriculture: Technology and economic perspectives*, 21-77.

Diederen, P., Van Meijl, H., Wolters, A., & Bijak, K. (2003). Innovation adoption in agriculture: Innovators, early adopters and laggards. *Cahiers d'Economie et de Sociologie Rurales, 67*, 29–50.

Eissa, M. (2024). Precision agriculture using artificial intelligence and robotics. *Journal of Research in Agriculture and Food Sciences, 1*(2), 35–35.

Javaid, M., Haleem, A., Khan, I. H., & Suman, R. (2023). Understanding the potential applications of Artificial Intelligence in Agriculture Sector. *Advanced Agrochem, 2*(1), 15–30.

Kumari, N., Chaudhary, D., Kaur, H., & Yadav, A. L. (2023). Artificial intelligence in supply chain optimization. In *2023 International Conference on IoT, Communication and Automation Technology (ICICAT)* (pp. 1-6). IEEE.

Mohammed, M., Hamdoun, H., & Sagheer, A. (2023). Toward sustainable farming: Implementing artificial intelligence to predict optimum water and energy requirements for sensor-based micro irrigation systems powered by solar PV. *Agronomy (Basel), 13*(4), 1081.

Nweje, U., & Taiwo, M. (2025). Leveraging Artificial Intelligence for predictive supply chain management, focus on how AI-driven tools are revolutionizing demand forecasting and inventory optimization. *International Journal of Science and Research Archive, 14*(1), 230–250.

Onteddu, A. R., Kundavaram, R. R., Kamisetty, A., Gummadi, J. C. S., & Manikyala, A. (2025). Enhancing Agricultural Efficiency with Robotics and AI-Powered Autonomous Farming Systems. *Malaysian Journal of Medical and Biological Research, 12*(1), 7–22.

Prajapati, C. S., Priya, N. K., Bishnoi, S., Vishwakarma, S. K., Buvaneswari, K., Shastri, S., Tripathi, S., & Jadhav, A. (2025). The Role of Participatory Approaches in Modern Agricultural Extension: Bridging Knowledge Gaps for Sustainable Farming Practices. *Journal of Experimental Agriculture International, 47*(2), 204–222.

Romeiro, A. S. A. (2022). *Technology assessment for sustainability in water use operationalization of a responsible governance based in responsible research and innovation (anticipation and inclusiveness)* (Doctoral dissertation, Universidade NOVA de Lisboa (Portugal)).

KEY TERMS AND DEFINITIONS

Automated Tasks: AI can automate tasks like weeding, harvesting, and sorting, freeing up farmers for more strategic decision-making and reducing labor costs.

Enhanced Decision-Making: AI provides data-driven insights for optimizing planting schedules, irrigation strategies, and other crucial decisions.

Improved Crop Health Monitoring: AI-powered image recognition and computer vision systems can detect diseases and pests early, allowing for timely interventions and preventing widespread crop losses.

Increased Efficiency and Productivity: AI-driven automation and optimized resource use lead to higher crop yields and greater overall farm efficiency.

Precision Farming: AI algorithms analyze vast amounts of data (soil conditions, weather patterns, crop health, etc.) to provide farmers with precise insights for resource management, leading to optimized fertilizer use, reduced water consumption, and minimized pesticide applications.

Sustainable Practices: By promoting precision farming and resource optimization, AI helps reduce environmental impact and contributes to more sustainable agricultural practices.

Chapter 3
The AI-Assisted Healthy Food Production

P. Selvakumar
https://orcid.org/0000-0002-3650-4548
Department of Science and Humanities, Nehru Institute of Technology, Coimbatore, India

H. Vidyalakshmi
https://orcid.org/0009-0005-1758-9637
JSS Academy of Higher Education and Research, India

R. Josphineleela
https://orcid.org/0000-0002-6250-9828
Panimalar Engineering College, India

V. Vaissnave
https://orcid.org/0000-0001-7333-531X
SRM Institute of Science and Technology, India

T. Ragupathi
https://orcid.org/0000-0002-2482-5096
SRM Institute of Science and Technology, India

Raj Kumar
Maharshi Dayanand University, Rohtak, India

T. C. Manjunath
https://orcid.org/0000-0003-2545-9160
Rajarajeswari College of Engineering, India

ABSTRACT

In decades, technological innovation has profoundly reshaped industries across the globe, and agriculture—long rooted in traditional methods—is no exception. transformative powerful catalyst for change, driving a new

DOI: 10.4018/979-8-3373-4827-8.ch003

era in food production that emphasizes efficiency, sustainability, and precision. As the global population steadily climbs toward by over 50%, (FAO). Meeting this demand in a context of finite arable land, climate volatility, labor shortages, and growing environmental concerns requires a rethinking of conventional agricultural practices. AI with learn from patterns, and make informed decisions, agricultural role in food production spans a wide spectrum—from crop selection and planting to harvesting, packaging, and distribution. This approach leverages tools like machine learning, optimize resource use, reduce waste, and determine, predict crop diseases, monitor weather conditions in real time.

INTRODUCTION: THE ROLE OF AI IN REVOLUTIONIZING FOOD PRODUCTION

In decades, technological innovation has profoundly reshaped industries across the globe, and agriculture—long rooted in traditional methods—is no exception. transformative powerful catalyst for change, driving a new era in food production that emphasizes efficiency, sustainability, and precision. As the global population steadily climbs toward by over 50%, (FAO). Meeting this demand in a context of finite arable land, climate volatility, labor shortages, and growing environmental concerns requires a rethinking of conventional agricultural practices. AI with learn from patterns, and make informed decisions, agricultural role in food production spans a wide spectrum—from crop selection and planting to harvesting, packaging, and distribution. At its core, AI facilitates agriculture, "precision agriculture." This approach leverages tools like machine learning, optimize resource use, reduce waste, and determine, predict crop diseases, monitor weather conditions in real time. These insights enable to make decisions, adjusting fertilizers with pinpoint accuracy, thereby the field, AI-powered drones and autonomous machinery have revolutionized traditional labor-intensive practices. These technologies can sow seeds, spray pesticides, and even harvest labor. Not only does this alleviate the burden on an increasingly aging and shrinking agricultural workforce, but it also ensures consistency in operations, leading to higher quality and more reliable outputs. Additionally, AI systems are capable of continual learning, improving their performance over time as they accumulate more data and experience. This foresight is critical in reducing risks and enabling supply chains to adapt more quickly to disruptions. For instance, if a specific region is likely to experience a drought, AI can help farmers plan alternative crop strategies or shift resources to mitigate losses. Similarly, during unexpected events like pandemics or geopolitical tensions, AI helps optimize food distribution and manage supply and demand imbalances. AI is also instrumental beyond the farm, in the post-harvest stages of food production. can detect imperfections or contamination in products using computer vision. These systems are more efficient and accurate than human inspectors, ensuring safety standards are met while reducing waste. AI is also used in packaging to extend shelf life and reduce environmental impact. For example, smart packaging that changes color based on freshness levels is increasingly informed by AI-based monitoring systems. Furthermore, AI technologies are being integrated into the logistics and retail supply chain. Smart systems help retailers predict customer demand, adjust stock levels, and reduce food waste. purchasing holidays or weather, these systems optimize supply schedules and ensure timely delivery. AI-powered chatbots and virtual

The Importance of Healthy Food Production for a Sustainable Future

In decades, technological innovation has profoundly reshaped industries across the globe, and agriculture—long rooted in traditional methods—is no exception. transformative powerful catalyst for change, driving a new era in food production that emphasizes efficiency, sustainability, and precision. (Prasad R. et al. 2021) As the global population steadily climbs toward by over 50%, (FAO). Meeting this demand in a context of finite arable land, climate volatility, labor shortages, and growing environmental concerns requires a rethinking of conventional agricultural practices. AI with learn from patterns, and make informed decisions, agricultural role in food production spans a wide spectrum—from crop selection and planting to harvesting, packaging, and distribution. (McCarthy, 2007) At its core, AI facilitates agriculture, "precision agriculture." This approach leverages tools like machine learning, optimize resource use, reduce waste, and determine, predict crop diseases, monitor weather conditions in real time. These insights enable to make decisions, adjusting fertilizers with pinpoint accuracy, thereby the field, AI-powered drones and autonomous machinery have revolutionized traditional labor-intensive practices. These technologies can sow seeds, spray pesticides, and even harvest labor. Not only does this alleviate the burden on an increasingly aging and shrinking agricultural workforce, but it also ensures consistency in operations, leading to higher quality and more reliable outputs. Additionally, AI systems are capable of continual learning, improving their performance over time as they accumulate more data and experience.

This foresight is critical in reducing risks and enabling supply chains to adapt more quickly to disruptions. For instance, if a specific region is likely to experience a drought, AI can help farmers plan alternative crop strategies or shift resources to mitigate losses. Similarly, during unexpected events like pandemics or geopolitical tensions, AI helps optimize food distribution and manage supply and demand imbalances. AI is also instrumental beyond the farm, in the post-harvest stages of food production. can detect imperfections or contamination in products using computer vision. (Mondal, 2020) These systems are more efficient and accurate than human inspectors, ensuring safety standards are met while reducing waste. AI is also used in packaging to extend shelf life and reduce environmental impact. For example, smart packaging that changes color based on freshness levels is increasingly informed by AI-based monitoring systems. (Xu Y. et al. 2021) Furthermore, AI technologies are being integrated into the logistics and retail supply chain. Smart systems help retailers predict customer demand, adjust stock levels, and reduce food waste. purchasing holidays or weather, these systems optimize supply schedules and ensure timely delivery. AI-powered chatbots and virtual assistants also enhance consumer engagement, offering personalized meal recommendations, nutritional advice, or even real-time updates on food origin and traceability. Sustainability is a core driver behind the adoption of AI in food production. With growing public concern over instance, precision agriculture significantly reduces water and chemical usage by applying inputs only where they are needed. Moreover, AI helps identify the most sustainable farming practices by simulating different scenarios and assessing their long-term ecological impact.

How AI Can Enhance Food Quality and Safety: Discussion of AI's impact on food quality.

Planet for future generations, production a sustainable future has never been more urgent. Sustainable food production encompasses the creation, processing, distribution, and consumption of maintains ecological balance, supports human health, promotes economic viability, upholds social. In this context, "healthy food" goes beyond mere nutritional value—it also implies food that is produced using environmentally responsible practices, free from harmful chemicals, accessible to all populations, and resilient to the shocks of climate change, pandemics, and economic disruption. (Zheng ZY. et al. 2017) As we approach a global population of nearly 10 billion by 2050, mitigating climate change, and fostering long-term planetary health. fossil fuels have enabled high yields but at a significant ecological cost. These practices degrade soil health, pollute water sources, and reduce ecosystem services that are vital for long-term productivity. Sustainable food production aims to reverse these trends by promoting methods that preserve natural resources, reduce carbon footprints, and enhance the resilience of ecosystems. (Sak J. et al. 2021) Techniques such as organic farming, regenerative agriculture, agroecology, and permaculture are increasingly recognized as pathways to produce food that nourishes people while protecting the environment. Healthy food production also addresses the critical issue of diet-related diseases and malnutrition. Modern diets, especially in urban and industrialized societies, are often high in ultra. This nutritional imbalance has led to a global epidemic of obesity, diabetes, cardiovascular diseases, and other chronic conditions, while many regions still suffer from undernutrition and micronutrient deficiencies. Sustainable food systems aim to make diverse, nutrient-rich, minimally processed foods more accessible and affordable to all people, regardless of socioeconomic status. By supporting local food systems, traditional diets, and diversified crop production, sustainable agriculture contributes to better health outcomes and more resilient communities.

One of the cornerstones. (Morgenstern JD. et al. 2021) Water, soil, and biodiversity are foundational to agriculture, and their conservation is essential for future food security. Sustainable practices emphasize, soil conservation techniques like cover cropping and no-till farming, and biodiversity enhancement through crop rotation and intercropping. These methods improve the long-term fertility of land, reduce the need for chemical inputs, and create habitats for beneficial insects and pollinators. In addition, the reduction of food waste across and storage retail and a critical component of sustainability. roughly moral failure in a world where hunger persists but also a massive loss of natural resources. Social equity and economic viability are equally vital pillars of sustainable food production. (Mazzei A. et al. 2015) A truly sustainable system must farmers and agricultural laborers, many of whom face exploitation and poverty despite being. Support for smallholder family, is especially important this regard. By strengthening local markets, improving access to credit and education, and encouraging cooperative models, sustainable food production can empower communities, reduce rural poverty, and enhance economic resilience. Additionally, inclusive policies that prioritize gender equity, land rights, and access to resources are essential food systems. The shift toward healthy and sustainable food production is also critical. Agriculture not only contributes significantly to global emissions but is also highly vulnerable to heatwaves, and shifting growing seasons. Sustainable practices that enhance soil carbon sequestration, reduce methane emissions from livestock, and promote agroforestry can play a pivotal role in reducing agriculture's carbon footprint. Moreover, resilient farming systems that integrate diverse crops, water-efficient methods, and community-based resource management are better equipped to withstand climatic shocks and ensure continuous food supply.

Precision Agriculture: AI-Powered Crop Monitoring and Management

In an increasingly interconnected and consumer-driven global food market, not just a regulatory requirement—it is a public health imperative and a competitive necessity. With growing concerns over foodborne illnesses, contamination, spoilage, adulteration, and ethical sourcing, modern food systems face intense scrutiny. Traditional methods of quality control, while foundational, are often reactive, labor-intensive, and limited in their scope. This is where emerges computer vision data analytics robotics—are reshaping how food is produced, processed, monitored, and delivered.

(Davenport T. et al. 2019) Their impact on food quality and safety is profound, offering real-time, predictive, and adaptive solutions that promise not only to enhance public health and consumer trust but also to streamline operations, reduce waste, and ensure compliance across complex supply chains. Computer vision systems equipped with high-resolution cameras and AI algorithms are increasingly used on food production lines to detect defects, foreign objects, discoloration, or textural abnormalities. These systems can identify anomalies that are often invisible to the human eye and do so with remarkable speed and precision. (Subramanian M. et al. 2020) For example, in fruit sorting, AI can distinguish between ripe and spoiled produce based on subtle color variations, surface texture, and shape. In meat processing plants, AI-powered cameras can detect contamination or improper cuts, ensuring uniformity and compliance with hygiene standards. This level of automation not only enhances quality increases throughput. Beyond visual inspection, predictive analytics for food safety. lab tests, environmental conditions, potential risks. For instance, likelihood microbial contamination-based temperature, humidity, handling procedures, and timelines. predictive producers take proactive measures, such as adjusting storage conditions, improving sanitation protocols, or even halting a batch before it reaches consumers. In the dairy industry, AI systems can monitor changes in milk composition and processing parameters to prevent spoilage and ensure consistent quality. These interventions not only protect consumers but also minimize costly recalls and brand damage.

AI-integrated sensors and IoT devices are revolutionizing how food quality is monitored across the supply chain. Smart sensors embedded in storage facilities, transport containers, and retail shelves can track temperature, humidity, gas levels (like ethylene or carbon dioxide), and even vibration during transportation. (Singh J. et al. 2023) These devices continuously transmit data to AI systems that analyze conditions in real time, issuing alerts when deviations from ideal parameters occur. For example, in cold chain logistics, AI can detect when refrigeration fails or temperatures exceed safe thresholds, allowing for immediate corrective action. This is particularly important for perishable goods like seafood, meat, and dairy, where even minor deviations can compromise safety and lead to significant losses. AI is also enhancing which are key factors in both food safety and consumer confidence. Blockchain integrated with AI allows every food journey—be recorded, verified, and tracked. (Dubey AK. et al. 2023) When a food safety issue arises, such as contamination or mislabeling, AI algorithms can rapidly analyze supply chain data to trace the source of the problem, identify affected batches, and isolate them efficiently. This minimizes the scope of recalls and prevents wider public exposure. Additionally, AI-powered platforms can accessible, handling, composition consume, thus greater accountability food industry. In the realm of food formulation and processing, AI is making strides in improving both quality and nutritional value. Using vast databases of ingredient properties, sensory profiles, and consumer preferences, AI systems can assist food scientists in designing new products that are not only tasty and appealing but also healthy and sustainable. (Côté M. et al. 2022) For example, AI can help reduce sugar or sodium in processed foods without compromising flavor, or suggest plant-based substitutes that replicate the texture and taste of animal-derived ingredients. This innovation contributes to better public health outcomes while growing demand, functional, ethically sourced products.

Predictive Analytics for Crop Yield and Disease Management

Increasing food demand, variability, limitations, agriculture undergoing a profound transformation. One of the most significant innovations driving this shift is approach leverages Artificial Intelligence (AI) and related technologies to optimize every aspect of productivity, promote sustainability crops. (Pesapane F. et al. 2018) AI, as the backbone of this revolution, empowers farmers with insights that were previously unimaginable, helping them understand intricate patterns in soil health, weather, pest behavior, irrigation, and yield potential. This paradigm shift from traditional farming to smart, precise, and predictive agriculture marks a new era where technology meets sustainability in the most impactful way. At its core, precision agriculture is about managing in ensure that crops receive exactly what they need—no more, no less. This level of granularity is made possible by integrating AI with a suite of digital tools such as GPS, drones, satellite imagery, remote sensing, IoT sensors, and robotics. Together, these technologies generate vast quantities of, plant and farming operations. AI systems then analyze this data using machine learning algorithms detect, forecast recommend targeted interventions. For example, by using AI to assess real-time aerial imagery captured by drones, farmers suffering from nutrient deficiencies or water stress and apply fertilizers or irrigation precisely improves health and reduces input costs.

Coupled with weather forecasting models, AI can also predict the optimal times for planting, irrigation, and harvesting. (Sharma N. et al. 2021) These forecasts, tailored to specific microclimates within a single field, enhance operational planning and reduce losses due to unpredictable weather events. For instance, AI systems can recommend delaying irrigation if rainfall is imminent or advise immediate pest control if conditions are ripe for an outbreak, thereby preserving yield quality and reducing unnecessary chemical usage. (Cordeiro F. et al. 2015) AI-driven soil analysis is another cornerstone of precision agriculture. Soil health is fundamental to sustainable crop production, and traditional testing methods are often slow, expensive, and infrequent. (Zepeda L. et al. 2008) AI enables real-time soil monitoring through sensors that measure pH levels, salinity, nutrient content, and microbial activity. These data points are processed using AI models that provide insights into soil fertility, recommend crop rotations, and suggest organic amendments. Over time, can offer increasingly precise prescriptions, helping maintain or restore soil vitality while minimizing dependence on synthetic fertilizers. This fosters long-term sustainability and reduces agriculture's contribution to environmental degradation.

Automated Farming: AI-Controlled Irrigation and Fertilization

The global agricultural landscape is undergoing a dramatic transformation, largely fueled by most revolutionary innovations in this space is predictive analytics, a branch of data science that harnesses the context of agriculture, predictive analytics pivotal enhancing yield managing plant diseases, two of the most crucial factors influencing food security, farmer profitability, and sustainable resource use. (Owens, 2023) By applying algorithms, machine learning models, and statistical techniques to vast datasets, predictive analytics allows farmers and agronomists to make informed decisions that reduce uncertainty, improve planning, and prevent costly losses due to unforeseen issues like weather extremes, pest outbreaks, or soil degradation. At its core, predictive analytics works integrating diverse types climatic, geospatial, agronomic, genetic, and sensor-based—to identify patterns and eye. (Ciocca G. et al. 2017) These data are sourced from satellites, drones, weather stations, IoT devices, soil sensors, historical yield records, and even market trends. The ability of predictive analytics to process this data and generate actionable insights is transforming farming from a reactive endeavor into a proactive and anticipatory science. For example, farmers can now receive predictions about expected yield weeks or months before harvest, allowing them to plan logistics, manage inputs efficiently, and make marketing decisions with confidence. would rely on experience, intuition, and basic metrics to estimate their yield, often resulting in inaccurate or overly optimistic predictions. Today, with predictive models, yield forecasting has become far more precise. (Vasiloglou MF. et al. 2020) These models use variables such as soil nutrient levels, crop variety, planting date, precipitation, temperature trends, and past performance data to simulate expected yields. For instance, a machine learning algorithm might data to assess vegetation health during various growth stages and project the final yield. This allows farmers to adjust their practices mid-season if forecasts show underperformance, such as increasing irrigation or supplementing with nutrients. Governments and agri-businesses also benefit from macro-level yield predictions, which inform national food security strategies, trade policies, and insurance schemes.

Beyond yield prediction, disease and pest outbreak forecasting is another game-changing area of application. Plant diseases and pests cause billions of dollars in losses annually and threaten the stability of food systems. Predictive analytics models can analyze environmental conditions such as humidity, temperature, rainfall, and wind patterns, along with crop phenology, to estimate the risk of disease emergence or pest infestation. For example, if weather forecasts predict a prolonged humid period during the flowering stage of wheat, a predictive model might signal a high risk of Fusarium head blight, enabling farmers to apply fungicides at the most effective time. (Huang Q. et al. 2018) Similarly, pest models can track the migration of insects like locusts or armyworms and alert farmers and authorities before outbreaks occur. This early warning capability not only reduces crop losses but also minimizes the need for blanket pesticide application, which is both costly and environmentally damaging. The success of predictive analytics relies heavily on come into play. (Hedrick VE. et al. 2012) Satellites and drones provide consistent and broad-scale imagery that captures vegetation health, land cover, and changes over time. IoT sensors placed in fields continuously monitor content, feeding real-time data into predictive systems. Mobile applications also enable farmers to input field observations—such as visual signs of disease—which enhance the accuracy of localized predictions. The aggregation of this data into centralized platforms allows for robust model training, validation, and continuous improvement over time.

AI-Powered Food Inspection and Quality Control: Analysis of food inspection.

Agriculture, one of the oldest human endeavors, is undergoing a profound transformation in the digital age. With the ever-increasing pressure to feed a growing global population, address climate change, conserve water, and reduce environmental degradation, traditional farming practices are no longer sufficient. As a result, automated farming, also known as smart or digital farming, has emerged as a technological breakthrough that revolutionizes the way food is produced. (Afreen S. et al. 2019) At the heart of this innovation are Artificial Intelligence (AI)-driven systems that control irrigation and fertilization—two of the most critical and resource-intensive operations in crop production. (Fakoor R. et al. 2019) These automated systems are enhancing productivity, reducing input waste, improving crop health, and enabling sustainable agricultural practices at scale. (Chib S. et al. 2025) AI-controlled irrigation and fertilization systems use a network manage farm inputs with extreme precision. Traditional irrigation and fertilization practices often rely on fixed schedules and manual applications, which can lead to overuse, runoff, soil degradation, or under-application that harms crop yield. AI automation replaces these inefficient methods with intelligent, responsive systems that adapt in real time to changing water and nutrients are delivered to crops exactly when, where, and in the amount needed—nothing more, nothing less. (Kaur M. et al. 2025) environmental and crop-specific data to optimize water use. Moisture sensors embedded in the soil collect continuous data on water availability at various depths, while weather stations and satellite systems provide updates on rainfall, evaporation rates, wind, and solar radiation. AI algorithms analyze this complex web of information to make accurate, localized decisions about when and how much to irrigate. (Kommineni M. et al. 2025) For instance, if a specific field section shows adequate moisture levels and weather forecasts predict rain, the AI system can delay or cancel irrigation, conserving water and reducing energy costs. Conversely, during dry spells, the system can ensure timely and targeted watering that maximizes absorption and minimizes waste through evaporation or deep percolation.

Figure 1. Revolutionizing the food industry

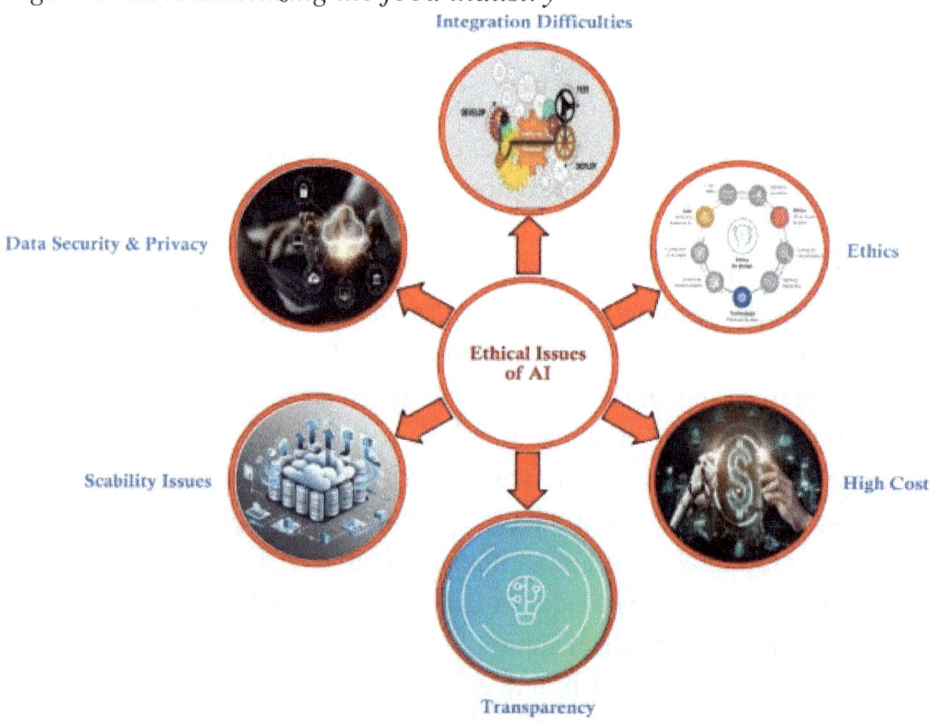

This intelligent water management is particularly water scarcity, which threaten agriculture globally. (Limbore NV. et al. 2025) In regions dependent on unpredictable rainfall or limited water sources, AI-controlled irrigation can ensure that every drop of water is used efficiently. Technologies such as drip irrigation and center-pivot irrigation have already reduced water consumption, but when enhanced by AI, their efficiency is multiplied. (Mazzei A. et al. 2015) AI systems can adjust pressure, flow rate, and distribution based on crop stage, time of day, and zone-specific soil characteristics. (McCarthy, 2007) This leads not only to water conservation but also to healthier plants and increased resilience during drought conditions. Similarly, AI-controlled fertilization, or fertigation when combined with irrigation, is revolutionizing nutrient management in agriculture. Crops require a balanced mix and potassium at different growth stages. Conventional fertilization often involves uniform application across large fields, resulting in nutrient imbalances, leaching, and pollution of water bodies. AI-powered systems overcome this by using data from soil sensors, plant tissue analysis, satellite imagery, and yield maps to determine the precise nutrient requirements of different field zones. (Mondal, 2020) Fertilizers are then automatically delivered via irrigation systems or specialized equipment in tailored doses, ensuring optimal nutrient availability without excess. These AI models also learn over time. (Morgenstern JD. et al. 2021) Through continuous monitoring and data collection, they refine their predictions and recommendations, adjusting to seasonal changes, soil depletion, or crop rotation practices. For example, if a particular area of a field consistently yields lower productivity, the system can analyze underlying factors such as soil compaction, pH imbalance, or micronutrient deficiency and recommend specific corrective actions. This dynamic learning capability not only improves fertilization outcomes but also supports long-term soil health and sustainability.

Smart Packaging: AI-Enabled Tracking and Monitoring

In a global food industry that spans continents and serves billions, maintaining consistent standards of quality and safety is both a monumental challenge and a non-negotiable necessity. From farm to fork, food products undergo a complex journey involving multiple stages—harvesting, processing, packaging, transportation, and retail—each posing unique risks to quality and safety. Traditional methods of food inspection, while critical, for food systems. Manual inspection processes are often labor-intensive, subjective, error-prone, and unable to keep pace with high-volume production. Enter transformative force how food inspection quality control is conducted, offering unprecedented levels of accuracy, speed, scalability, and reliability. (Pesapane F. et al. 2018) AI-powered inspection systems are reshaping the food industry by enhancing real-time decision-making, reducing waste, ensuring compliance with stringent regulations, and most importantly, protecting consumer health. At the heart of AI-driven food inspection are machine learning (ML) algorithms and computer vision systems that enable machines to "see," analyze, and evaluate food products with extraordinary precision. (Prasad R. et al. 2021) These systems utilize high-resolution cameras, multispectral and hyperspectral imaging, thermal scanners, and X-ray sensors to capture visual and structural data about food items. (Sak J. et al. 2021) AI algorithms, trained on vast datasets of product images and inspection criteria, can then identify defects, contaminants, inconsistencies, or abnormalities in size, shape, color, texture, or composition. (Samal A. et al. 2025) For instance, in a fruit processing plant, computer vision systems can sort apples based on ripeness, detect bruises or mold, and remove damaged items from the production line—all in real time and with minimal human intervention.

One of the most transformative aspects of AI-powered inspection is its speed and scalability. Human inspectors can only evaluate a limited number of products within a given time frame, and their judgment may be influenced by fatigue, bias, or inconsistency. (Sharma N. et al. 2021) AI systems, on the other hand, can inspect thousands of items per minute with consistent criteria and zero fatigue. This high-throughput capability is critical in large-scale operations such as poultry processing or beverage bottling, where even brief delays or undetected issues can result in massive financial and reputational losses. Moreover, AI systems can operate continuously, enabling 24/7 quality monitoring contaminated reaching consumers. AI also plays a vital role in foreign object detection (FOD), one of the most serious hazards in food production. (Singh J. et al. 2023) Using technologies like X-ray inspection, hyperspectral imaging, and ultrasonic sensors, AI can detect non-food materials such as plastic, glass, metal, or bone fragments that may inadvertently enter the food supply during processing. These systems are not only more sensitive than human inspection but also capable of differentiating between organic and inorganic materials with a high degree of accuracy meat, identify tiny bone shards in fillets that are invisible to the naked eye, ensuring compliance with regulatory standards and preventing consumer injury. Another crucial application is in contaminant and allergen detection. AI-powered systems can analyze chemical, microbiological, and molecular signatures using spectroscopic methods to detect harmful substances, or allergens like peanuts and gluten. Traditional lab testing is often time-consuming and reactive, with results available only after processing is completed. (Soundariya A. et al. 2025) AI brings the capability of real-time, non-invasive testing, allowing food producers to identify contamination early and take immediate corrective action. This reduces the incidence of recalls, which can be devastating for both public health and brand reputation. Beyond safety, AI also enhances aesthetic and sensory quality assessment, which is essential for meeting consumer expectations. From the perfect golden hue of potato chips to the uniform thickness of chocolate bars, consumers associate visual appeal with quality. AI systems can assess these attributes consistently and objectively. In bakeries, for example, AI can measure crust color, loaf height, and texture to ensure every batch meets predefined quality standards. In beverage bottling, systems can detect improper sealing, fill levels, or label misalignment—details that, while seemingly minor, have a major impact on consumer perception and trust.

Optimizing Food Processing: AI-Driven Supply Chain Management

As the global food supply chain becomes more intricate and consumer expectations for safety, transparency, and sustainability grow, packaging is evolving far beyond its traditional role of merely containing and protecting products. Enter smart packaging, a transformative innovation that combines Artificial Intelligence (AI), sensors, data analytics, and connectivity technologies to enable real-time tracking, monitoring, and communication across every stage of a product's life cycle. Smart packaging is not only revolutionizing how food is stored, transported, and consumed, but also AI into packaging systems allows data to be captured, analyzed, and acted upon autonomously, marking a significant shift from passive packaging to dynamic, intelligent packaging solutions. (Xu Y. et al. 2021) At the core of AI-enabled smart packaging is the ability to sense and communicate environmental and product-specific data. Smart packaging solutions often incorporate GPS modules, temperature sensors, humidity detectors, gas sensors, and biosensors. These components can be embedded into the packaging itself or applied externally, allowing products to transmit data regarding their location, storage conditions, and shelf life. for manufacturers, retailers, logistics providers, and even consumers. For instance, a shipment of fresh seafood equipped with smart packaging can monitor temperature levels throughout transport. If a temperature excursion occurs, AI can alert the supplier in real time, enabling immediate corrective action before spoilage occurs.

One of the most impactful applications of smart packaging is in cold chain management. Many perishable products, including dairy, meat, seafood, and pharmaceuticals, require strict temperature control to remain safe and effective. Traditional methods of cold chain monitoring involve periodic checks that leave significant gaps during which spoilage can occur unnoticed. Smart packaging, on the other hand, provides continuous temperature and humidity logging, with AI models predicting the product's remaining shelf life based on deviations from optimal conditions. If the temperature of a milk shipment rises above a critical threshold during transit, the system can forecast accelerated spoilage and notify distributors to reroute the shipment or halt distribution. (Zepeda L. et al. 2008) This minimizes both health risks and economic losses. AI also enables predictive analytics in smart packaging, allowing stakeholders to anticipate and prevent issues before they escalate. For example, by analyzing historical data from multiple batches of packaged produce, AI can detect patterns that indicate higher spoilage rates under certain logistical conditions. It can then recommend changes in packaging material, handling practices, or delivery routes. Moreover, AI can be used to model consumer demand, enabling dynamic shelf-life labeling that reflects real-time product freshness instead of static expiration dates. This approach helps retailers manage inventory more effectively and empowers consumers to make informed decisions, reducing the amount of food that is discarded due to conservative or misleading labeling.

Benefits of AI-Assisted Food Production: Increased Efficiency and Sustainability

Contemporary global economy, faces enormous challenge rapidly contending rising costs, supply chain disruptions, environmental concerns, and increasingly complex consumer demands. Traditional food supply chains—often linear, rigid, and fragmented—are being reimagined through. Among most impactful these technologies is which across the food processing sector. (Zheng ZY. et al. 2017) AI-driven supply chain systems enable unprecedented visibility, flexibility, and efficiency, allowing stakeholders to make smarter decisions, anticipate disruptions, reduce waste, and streamline the movement of goods from farm to fork. By processing identifying hidden generating predictive, AI transforming food supply chains into intelligent, responsive networks that support both business performance and global food security. Food processing is at the heart of the supply chain—it links agricultural production with consumption and involves converting raw materials into consumable products. This stage includes sorting, cleaning, cooking, packaging, and transporting food, and it is highly dependent on the synchronization of multiple upstream and downstream activities. critical role in optimizing these processes strategic planning tools that minimize inefficiencies and boost throughput. In traditional models, supply chain decisions are often reactive, based on historical data or subjective judgment. With AI, food processors can make proactive, data-driven decisions that adapt dynamically to changing conditions such as weather fluctuations, demand shifts, labor shortages, and geopolitical events. One of the primary ways AI enhances supply chain management is through minimizing food waste, reducing stockouts, and optimizing production schedules. (Soundariya A. et al. 2025).

AI algorithms analyze a wide range of variables including historical sales data, seasonal trends, weather forecasts, social media activity, and even global news to predict consumer demand with high accuracy. (Singh J. et al. 2023) For example, during a heatwave, AI may predict a spike in demand for cold beverages or fresh produce and adjust procurement, production, and distribution plans accordingly. (Sharma N. et al. 2021) This helps avoid overproduction, which can lead to spoilage, or underproduction, which causes lost sales and customer dissatisfaction. (Samal A. et al. 2025) AI also monitors real-time inventory levels and automatically triggers reorders or production shifts based on forecasted needs, ensuring a just-in-time inventory model that reduces waste and capital lock-in. Beyond forecasting, AI significantly improves logistics and transportation management, one of the most complex and cost-intensive elements of the food supply chain. (Sak J. et al. 2021) Transportation of perishable goods is highly time-sensitive and vulnerable to disruptions caused by traffic, road conditions, fuel shortages, or customs delays. AI-driven systems use GPS tracking, IoT sensors, and traffic data to optimize delivery routes, predict arrival times, and monitor vehicle conditions. These systems can dynamically reroute trucks in response to real-time obstacles, ensuring that products reach their destination within optimal freshness windows. Furthermore, temperature-controlled logistics (cold chain) can be managed using AI to continuously assess the integrity of goods in transit, sending alerts if temperature thresholds are breached. This real-time logistics intelligence minimizes spoilage, reduces operational costs, and supports regulatory compliance.

Challenges and Limitations of AI in Food Production: Discussion of challenges.

Artificial Intelligence (AI) is rapidly becoming one of the most transformative forces across the global food sector, driving unprecedented changes in robotics, and automation, AI is enhancing operational efficiency while significantly contributing to sustainability. From the farm to the processing plant, AI is reducing resource consumption, optimizing inputs, increasing yields, minimizing waste, and supporting environmentally conscious practices. This fusion of digital intelligence with agricultural and food technologies is not only revolutionizing productivity but also aligning the food system with broader goals of resilience, equity, and ecological balance. One of the most immediate and tangible food production dramatic increase operational efficiency. (Prasad R. et al. 2021) volumes generated from sensors, drones, satellite imagery, IoT devices, and historical records to guide real-time decision-making. On farms, AI supports precision agriculture by enabling exact water, crop- needs, soil, and weather forecasts. This eliminates the inefficiencies of blanket treatments, ensuring that resources are used optimally while reducing environmental runoff. (Pesapane F. et al. 2018) In greenhouses, AI can control climate, lighting, and irrigation automatically, adapting conditions to the specific growth stages of crops, thereby boosting productivity per square meter. In processing and manufacturing, workflows, production speed, labor dependency. Smart sorting machines equipped with computer vision can classify fruits, vegetables, and grains based on size, ripeness, and defects far more accurately than human workers. Robotics integrated with AI can manage repetitive tasks such as cutting, packaging, labeling, and palletizing with consistency and minimal error, leading to faster turnaround times and lower operational costs. Predictive maintenance, powered by AI, helps avoid unplanned downtimes by alerting managers to machinery issues. This ability to pre-emptively address inefficiencies contributes to maximum uptime and resource utilization, throughput profitability.

Optimizing supply chain logistics, which ensuring move efficiently production sites to consumers. AI can forecast demand more accurately by analyzing consumer behavior, market trends, weather conditions, and socio-economic data. With better demand planning, producers can avoid overproduction, reduce inventory waste, and ensure products reach markets just in time. (Morgenstern JD. et al. 2021) AI algorithms can also recommend optimal transportation routes, balancing speed, fuel efficiency, and environmental impact. For perishable goods, AI ensures that cold chain logistics maintain ideal conditions, preventing spoilage and maintaining food quality. (Mondal, 2020) These enhancements to supply chain dynamics mean fewer resources wasted, lower carbon emissions, and improved access to fresh food. In addition to operational efficiency, AI significantly contributes to resource sustainability—biggest resource challenges in agriculture are water usage, especially in arid and drought-prone regions. (McCarthy, 2007) AI-powered irrigation systems use real-time soil moisture data, plant health assessments, and weather predictions to irrigate only when and where necessary. (Mazzei A. et al. 2015) This has been shown to reduce water usage by 20–50% compared to traditional irrigation methods. Similarly, AI helps farmers apply fertilizers and pesticides more judiciously, using data models that determine the precise dosage and timing required for optimal plant health. This not only lowers input costs but also reduces chemical runoff into waterways, protecting ecosystems and public health. Energy efficiency is another area where AI demonstrates clear sustainability benefits. In food processing plants, AI can regulate machinery and climate control systems based on production needs and energy prices. Smart energy management systems can reduce peak electricity usage, integrate renewable energy sources, and optimize refrigeration cycles, thereby decreasing greenhouse gas emissions. For example, an AI system can analyze patterns of energy consumption and adjust operations to minimize electricity use during peak hours without compromising output. As global concerns about climate change intensify, these AI-enabled efficiencies are essential for helping the food industry meet emissions reduction targets.

Addressing Food Safety and Regulatory Compliance with AI: Overview of food safety.

While Artificial Intelligence (AI) holds immense promise for revolutionizing food production, its implementation is not without substantial challenges and limitations. Despite the significant progress made in integrating AI technologies across agriculture, food processing, logistics, and retail, numerous practical, economic, technical, ethical, and social issues still hinder its full-scale adoption. These obstacles must be critically examined to ensure that AI solutions in food production are not only effective and scalable but also inclusive, sustainable, and ethically sound. (Limbore NV. et al. 2025) A nuanced understanding of these challenges is essential for policymakers, technologists, agronomists, food industry leaders, and other stakeholders to strategically address barriers and build resilient AI-driven food systems. foremost challenges high adoption, particularly enterprises (SMEs) and farmers in developing regions. Implementing AI solutions often requires significant investment in advanced hardware (e.g., sensors, drones, robotics), software platforms, data infrastructure, and skilled personnel. Many farmers and food processors operate on thin margins and lack the capital needed to invest in cutting-edge technologies. (Kommineni M. et al. 2025).

Another major limitation is insufficient digital infrastructure, especially in rural or remote areas. (Kaur M. et al. 2025) AI systems depend on reliable internet connectivity, cloud computing, and power supplies to collect, transmit, and process data in real time. In many parts of the world, these infrastructures are either absent or unreliable. Even in well-developed regions, the bandwidth and latency required for high-resolution imaging, real-time sensor data, and autonomous machinery can strain existing systems. (Chib S. et al. 2025) Without the necessary digital backbone, the severely curtailed, making infrastructure development a prerequisite for widespread AI deployment. Closely tied to infrastructure is the challenge of and manual entries. However, many of these data sources are fragmented, inconsistent, or siloed within proprietary platforms. Additionally, historical data may be incomplete, biased, or unstandardized, making it difficult to train accurate and generalizable models. The lack of interoperable data formats and common standards further hampers the seamless integration of AI systems across the value chain. Poor data quality leads to poor predictions, which can result in incorrect decisions and operational inefficiencies. Technical limitations of AI models also pose significant challenges. While AI excels at pattern recognition and predictive modeling, it often lacks contextual understanding and adaptability. (Fakoor R. et al. 2019) Many AI systems perform well under controlled conditions but struggle in the unpredictable and variable environments typical of agriculture and food processing.

Future Trends in AI-Assisted Food Production: Vertical Farming and Robotics: Analysis of future trends.

Pillar, consumer trust, and within the global food system. meet stringent regulatory standards is a complex challenge that involves producers and processors distributors retailers. The increasing complexity of modern food supply chains—characterized by globalization, intricate product formulations, and diverse consumer preferences—has made food safety management more demanding than ever food safety protocols, automate compliance monitoring, predict potential hazards, and streamline regulatory adherence, thereby helping to protect consumers and strengthen the integrity of the food industry. (Afreen S. et al. 2019) At its core, food safety encompasses all interventions foodborne contamination. This includes managing microbial pathogens, chemical residues, allergens, physical contaminants, and spoilage throughout production, processing, storage, and distribution. worldwide—set rigorous guidelines, standards, and inspection protocols designed to mitigate risks and ensure product quality. Compliance with these regulations is mandatory for all food businesses and involves regular inspections, documentation, traceability, and recall mechanisms. However, traditional food safety management often relies heavily on manual inspections, retrospective data analysis, and fragmented record-keeping, which can lead to inefficiencies, delays, and gaps in risk detection. AI addresses many of these challenges by scan food products on production lines at speeds and accuracies far exceeding human inspectors. (Hedrick VE. et al. 2012) These systems detect anomalies such as discoloration, texture inconsistencies, foreign objects, and packaging defects that could indicate contamination or spoilage. For instance, AI-enabled cameras can distinguish bruised fruits or cracked eggs with high precision, frees up labor for more complex quality assurance activities.

Beyond surface-level inspection, AI supports microbial risk management through advanced sensor integration and data analysis. Sensors embedded in processing equipment, packaging, or storage environments continuously monitor parameters like temperature, humidity, and gas composition, which are critical for controlling microbial growth. this safe predict contamination events before they escalate identify patterns indicating early onset bacterial proliferation in refrigerated goods, triggering alerts to operators who can take corrective action. (Huang Q. et al. 2018) This proactive approach enhances food safety by shifting from reactive responses to preventive control. Traceability is another domain where AI significantly bolsters food safety and regulatory compliance. Modern supply chains involve numerous actors, often spanning multiple countries and jurisdictions. Ensuring full visibility into handling, food from farm inputs to final retail shelf. When a safety concern arises, AI can rapidly analyze traceability data to identify the source of contamination, pinpoint affected batches, and coordinate targeted recalls. This reduces the scope and cost of recalls while protecting public health. (Vasiloglou MF. et al. 2020) Additionally, AI can verify compliance with labeling regulations, allergen declarations, and expiration dates, further safeguarding consumers.

Opportunities for Small-Scale Farmers: AI-Powered Solutions

Evolution, driving innovations that promise to reshape how we grow, harvest, and deliver food. Among the most promising and rapidly advancing future trends in AI-assisted food production are vertical farming and robotics. These technologies, synergistically integrated with AI, are poised to address critical challenges such as land scarcity, climate variability, labor shortages, and environmental sustainability. (Ciocca G. et al. 2017) By robotic automation, vertical farming and robotics herald food production—one precision, resilience, and scalability. Vertical farming represents a radical departure from traditional agriculture, moving crop cultivation indoors and stacking production vertically in controlled environments. This approach offers the potential to grow food year-round, independent of weather or geographic constraints, thereby natural disasters vertical farming environmental controls, and crop management. (Owens, 2023) Through a combination of sensors, cameras, then analyze this data to create precise growth protocols tailored to each crop species and growth stage. AI's ability to dynamically adjust lighting spectra and intensity using LED technology, regulate irrigation with hydroponic or aeroponic systems, and maintain ideal atmospheric conditions leads to maximal photosynthetic efficiency and biomass production. This fine-tuned control not only enhances crop yields but also reduces field farming, while virtually eliminating pesticide application.

Vertical farms, powered by AI, also minimize land use, allowing urban centers to produce significant amounts of fresh produce locally, thereby slashing food miles and associated carbon emissions. This localized production addresses food security concerns and increases access to fresh, nutrient-rich foods, especially in densely populated cities or regions with poor arable land. (Zepeda L. et al. 2008) Robotics, intertwined with AI, forms the backbone of the automation revolution within vertical farms and broader agricultural systems. Automated robotic systems are increasingly capable of performing complex tasks such as seeding, transplanting, pruning, harvesting, and packaging with high precision and repeatability. AI-powered vision systems enable robots to identify plant health issues, detect diseases or nutrient deficiencies, and make nuanced decisions on which plants to prune or harvest, ensuring product quality and minimizing waste. (Cordeiro F. et al. 2015) These robotic platforms can operate 24/7, significantly reducing the reliance on human labor—a critical advantage given the widespread labor shortages and the physically demanding nature of agricultural work. Furthermore, AI-enabled robotic systems in vertical farming can seamlessly coordinate with environmental control systems, adapting farming operations to changing conditions without human intervention. For example, if AI detects a fungal infection developing in a crop section, robotic sprayers can precisely target the affected plants with appropriate treatments, reducing chemical use and preventing widespread contamination. This convergence of AI and robotics facilitates highly scalable, consistent, and sustainable food production that can be replicated in multiple locations worldwide.

The Role of AI in Achieving Sustainable Development Goals

The backbone of global agriculture, producing a substantial portion, especially developing countries. (Sharma N. et al. 2021) Yet, they often face, including limited access technology, capital, markets, and agricultural knowledge, which constrain productivity, income, and sustainability. The advent of unprecedented empower smallholder farmers by providing accessible, affordable, and scalable solutions tailored to their specific needs. AI-powered technologies can enhance decision-making, optimize resource use, improve crop management, increase resilience against climate variability, and open new market avenues. By bridging the technological divide, AI can catalyze rural development, reduce poverty, and contribute to global food security. One of the most transformative opportunities AI offers small-scale farmers is personalized agronomic advisory services delivered through mobile phones and low-cost devices. Many farmers lack timely and accurate information on weather forecasts, soil health, pest and disease threats, satellite imagery, sensor data, and historical patterns—to provide tailored recommendations. For example, an AI-powered app can advise a farmer on the optimal planting time, irrigation schedule, or fertilizer application rate based on their specific field conditions and crop type. This precision farming approach helps smallholders maximize yields while minimizing input costs and environmental impact. AI's role in early warning systems for pest and disease outbreaks is particularly valuable for small-scale farmers who might lack access to traditional extension services.

Moreover, AI facilitates resource optimization, helping small-scale farmers use water, fertilizers, and pesticides more efficiently. Given that these inputs often constitute the largest share of production costs, AI tools that recommend precise quantities status, moisture growth stage can significantly boost profitability. For instance, AI-controlled irrigation systems activated can reduce. Such systems are, where efficient resource management enhances sustainability and resilience. Access to market information and financial services is another key opportunity enabled by AI. (Pesapane F. et al. 2018) access. AI platforms aggregate data on commodity prices, demand trends, and buyer networks, connecting farmers directly to local or international markets, thereby improving price transparency and reducing exploitation by intermediaries. Furthermore, AI-powered credit scoring models mobile history, farm performance, enabling smallholders to secure loans and insurance products tailored to their risk profiles. This financial inclusion supports investment in improved inputs, technology adoption, and climate risk mitigation. In addition, AI-powered supply chain solutions help small-scale farmers organize collectively to reduce post-harvest losses, improve storage, and streamline distribution. AI optimizes logistics by predicting demand, routing delivery vehicles efficiently, and coordinating harvest schedules, thereby enhancing freshness and reducing waste.

Harnessing the Power of AI for a Healthier and More Sustainable Food System

Combating climate change. Achieving these ambitious objectives requires innovative, scalable, and efficient approaches that leverage the potential of emerging technologies. Artificial Intelligence (AI), with its unparalleled capabilities in data processing, prediction, optimization, and automation, is increasingly recognized as a pivotal enabler in accelerating progress towards the SDGs. By transforming industries, improving decision-making, and enhancing resource management, AI can help overcome complex development in multiple dimensions. One of the fundamental ways AI supports the SDGs is by enhancing data availability and analysis, which is crucial for informed policymaking and monitoring progress. Many SDGs suffer from data gaps and reporting delays, particularly in low-income countries. AI sensor networks, social media, and administrative records to generate real-time insights on environmental changes, population health, economic activity, and infrastructure status. For example, AI-powered remote sensing is used to monitor deforestation, water quality, and crop health, providing accurate and timely information that informs conservation strategies and agricultural policies. These data-driven insights enable governments, organizations, and communities to allocate resources efficiently, identify vulnerable populations, and track SDG indicators more effectively. AI also contributes significantly to ending hunger (SDG 2) and promoting sustainable agriculture by optimizing food production systems. (Subramanian M. et al. 2020) smallholder with tailored recommendations and improving supply chain logistics, AI helps.

In the realm of health and well-being (SDG 3), AI is transforming resource allocation for emergency responses. Telemedicine platforms enhanced with AI facilitate remote consultations and health monitoring, overcoming geographical and infrastructural barriers. AI's potential to improve health outcomes is critical for achieving universal health coverage and reducing mortality rates. optimizing water management and quality control. Intelligent sensors combined with AI algorithms monitor water distribution networks for leaks and contamination, reducing waste and ensuring safe drinking water. Predictive analytics help manage water demand and supply, particularly in water-scarce regions. AI-driven models support wastewater treatment processes and help protect aquatic ecosystems, contributing to the sustainable management of water resources. Affordable and clean energy (SDG 7) benefits from AI's ability to optimize energy production, distribution, and consumption. efficiency systems by forecasting and power availability balancing grid loads. (Davenport T. et al. 2019) Smart grids powered by AI improve energy storage, demand response, and fault detection, increasing reliability and reducing carbon footprints efficient building transportation further contribute to reducing global greenhouse gas emissions. AI's impact on is profound, as it drives productivity gains, creates new markets, and fosters entrepreneurship. AI-powered automation and robotics increase manufacturing efficiency and reduce costs, while AI-enabled platforms support small and medium enterprises by providing market insights, financial services, and supply chain optimization. By democratizing access to knowledge and innovation, AI stimulates inclusive economic growth and job creation.

CONCLUSION

Food system stands at a critical juncture, confronted by mounting the urgent need for improved nutrition and food security. To navigate these complex issues, compelling pathway toward creating a food system that is healthier, more sustainable, and resilient. Future directions in AI-driven food production, processing, distribution, and consumption are poised to revolutionize the agricultural landscape, enabling smarter decision-making, efficient resource use, enhanced food quality, and reduced environmental impacts. By integrating AI technologies throughout the entire food value chain, we can move beyond incremental improvements to achieve transformative, systemic change viability stewardship. A foundational future direction lies in advancing AI-enabled precision agriculture, where AI-powered tools will increasingly deliver hyper-local, data-driven insights that optimize every facet of crop and livestock management. With the proliferation of affordable sensors, drones, satellite imagery, and IoT devices, AI algorithms will harness vast datasets to monitor soil health, moisture levels, nutrient status, pest populations, and crop growth patterns in real-time. Machine learning models will evolve to provide farmers with predictive analytics that anticipate weather extremes, disease outbreaks, and yield forecasts, enabling proactive interventions rather than reactive responses. This shift from generalized to highly customized farming practices will maximize productivity while minimizing freeing farmers to focus on strategic management and innovation. Enhancing food quality safety also significantly. Future AI systems will incorporate advanced computer vision and sensor technologies capable of detecting contamination, spoilage, and adulteration at multiple stages—on the farm, during processing, in packaging, and at points of sale. These systems will not only improve consumer safety but also reduce food waste by identifying produce that meets quality thresholds for various markets, optimizing supply chains accordingly. Blockchain and AI integration will enable transparent origin, handling, sustainability credentials in real-time, fostering trust and informed purchasing decisions. Additionally, AI-driven predictive maintenance and quality control in food processing facilities will enhance operational reliability. Storage conditions to dynamically optimize the flow of food from producers to consumers. By forecasting fluctuations in supply and demand with unprecedented accuracy, AI will facilitate just-in-time deliveries that minimize spoilage and overproduction. Autonomous vehicles and drones, guided by AI, will increasingly handle last-mile delivery in urban and rural areas, improving accessibility to fresh food, especially in underserved communities. Furthermore, AI-enabled platforms will support smallholder farmers and local producers by linking them directly to markets, creating more equitable value chains and supporting local economies. Nutrition and public health represent another frontier where AI will have profound impacts. Future AI applications will analyze dietary data, genetic profiles, and lifestyle factors to offer personalized nutrition recommendations that AI-powered apps will help consumers plan meals that optimize nutritional intake while minimizing environmental impact, guiding choices towards sustainably sourced foods with lower carbon footprints. Additionally, AI tools will assist policymakers in designing food systems and subsidies that encourage production and consumption patterns aligned with both health and sustainability goals, supporting global efforts to combat obesity, malnutrition, and food insecurity simultaneously.

REFERENCES

Chib, S., Mehta, A. K., Selvakumar, P., Saravanan, A., Mishra, B. R., & Manjunath, T. C. (2025). Mental Health Challenges and Solutions in High-Pressure Work Environments. In Harriott, S. (Ed.), *Innovative Approaches to Managing Conflict and Change in Diverse Work Environments* (pp. 247–272). IGI Global Scientific Publishing.

Ciocca, G., Napoletano, P., & Schettini, R. (2017). Food recognition: A new dataset, experiments, and results. *IEEE Journal of Biomedical and Health Informatics*, *21*(3), 588–598. PMID: 28114043

Cordeiro, F., Bales, E., Cherry, E., & Fogarty, J. (2015). Rethinking the mobile food journal: Exploring opportunities for lightweight photo-based capture. In *Proceedings of the 33rd Annual ACM Conference on Human Factors in Computing Systems* (pp. 3207–3216). ACM.

Côté, M., & Lamarche, B. (2022). Artificial intelligence in nutrition research: Perspectives on current and future applications. *Applied Physiology, Nutrition, and Metabolism*, *47*(1), 1–8. PMID: 34525321

Davenport, T., & Kalakota, R. (2019). The potential for artificial intelligence in healthcare. *Future Healthcare Journal*, *6*(2), 94–98. PMID: 31363513

Dubey, A. K., Chabert, G. L., Carriero, A., Pasche, A., Danna, P. S., & Agarwal, S. (2023). Ensemble deep learning derived from transfer learning for classification of COVID-19 patients on hybrid deep-learning-based lung segmentation: A data augmentation and balancing framework. *Diagnostics (Basel)*, *13*(11), 1954. PMID: 37296806

Kaur, M., Selvakumar, P., & Dinesh, N. Mohit, Chandel, P. S., & Manjunath, T. C. (2025). Reconstruction and Preservation Strategies for Cultural Heritage. In S. Oflazoğlu & M. Dora (Eds.), Preserving Cultural Heritage in Post-Disaster Urban Renewal (pp. 385-412). IGI Global Scientific Publishing.

Kommineni, M., Chundru, S., Maroju, P. K., & Selvakumar, P. (2025). Ethical Implications of AI in Sustainable Development Pedagogy. In Chinedu, C., Mafarja, N., Hussein, M., & Saleem, A. (Eds.), *Rethinking the Pedagogy of Sustainable Development in the AI Era* (pp. 17–36). IGI Global Scientific Publishing.

Limbore, N. V., Anute, N., Ahluwalia, G. K., Chordiya, P. B., Kalshetti, P., & Selvakumar, P. (2025). Ethical Leadership and Corporate Responsibility. In Harriott, S. (Ed.), *Innovative Approaches to Managing Conflict and Change in Diverse Work Environments* (pp. 193–218). IGI Global Scientific Publishing.

Mazzei, A., Anselma, L., De Michieli, F., Bolioli, A., Casu, M., Gerbrandy, J., & Lunardi, I. (2015). Mobile computing and artificial intelligence for diet management. In *New Trends in Image Analysis and Processing—ICIAP 2015 Workshops* (pp. 342–349). Springer International Publishing.

McCarthy, J. (2007). From here to human-level AI. *Artificial Intelligence*, *171*(18), 1174–1182.

Mondal, B. (2020). Artificial intelligence: state of the art. In Balas, V., Kumar, R., & Srivastava, R. (Eds.), *Recent trends and advances in artificial intelligence and internet of things* (pp. XX–XX). Springer.

Morgenstern, J. D., Rosella, L. C., Costa, A. P., De Souza, R. J., & Anderson, L. N. (2021). Perspective: Big data and machine learning could help advance nutritional epidemiology. *Advances in Nutrition*, *12*, 621–631. PMID: 33606879

Owens, S. J. (2023). The 8 best food tracker apps of 2023. Lifewire. Retrieved January 3, 2023, from Pesapane, F., Codari, M., & Sardanelli, F. (2018). Artificial intelligence in medical imaging: Threat or opportunity? European Radiology Experimental, 2, 35.

Prasad, R., & Choudhary, P. (2021). State-of-the-art of artificial intelligence. [Google Scholar.]. *Journal of Mobile Multimedia*, *17*, 427–454.

Sak, J., & Suchodolska, M. (2021). Artificial intelligence in nutrient science research: A review. *Nutrients*, *13*(2), 322. PMID: 33499405

Samal, A., & Misra, L. I. Mohit, Pougajendy, S., Chandel, P. S., Manjunath, T. C., & Selvakumar, P. (2025). Musealization and Tourism Practices. In S. Oflazoğlu & M. Dora (Eds.), Preserving Cultural Heritage in Post-Disaster Urban Renewal (pp. 115-142). IGI Global Scientific Publishing.

Sharma, N., Sharma, R., & Jindal, N. (2021). Machine learning and deep learning applications—A vision. *Global Transitions Proceedings*, *2*(1), 24–28.

Singh, J., Singh, N., Fouda, M. M., Saba, L., & Suri, J. S. (2023). Attention-enabled ensemble deep learning models and their validation for depression detection: A domain adoption paradigm. *Diagnostics (Basel)*, *13*(12), 2092. PMID: 37370987

Soundariya, A., Behera, N. K., Selvakumar, P., Chandel, P. S., Gupta, S., & Manjunath, T. C. (2025). AI and Environmental Ethics: The Role of AI in Addressing Climate Change Ethical Considerations in AI-Driven Environmental Solutions. In Chakraborty, S. (Ed.), *Human Values, Ethics, and Dignity in the Age of Artificial Intelligence* (pp. 321–346). IGI Global Scientific Publishing.

Subramanian, M., Wojtusciszyn, A., Favre, L., Boughorbel, S., Shan, J., Letaief, K. B., & Chouchane, L. (2020). Precision medicine in the era of artificial intelligence: Implications in chronic disease management. *Journal of Translational Medicine*, *18*(1), 1–12. PMID: 33298113

Vasiloglou, M. F., Christodoulidis, S., Reber, E., Stathopoulou, T., Lu, Y., & Stanga, Z. (2020). What healthcare professionals think of "nutrition & diet" apps: An international survey. *Nutrients*, *12*(8), 2214. PMID: 32722339

Xu, Y., Liu, X., Cao, X., Huang, C., Liu, E., Qian, S., & Zhang, J. (2021). Artificial intelligence: A powerful paradigm for scientific research. *Innovation*, *2*(4), 100179. PMID: 34877560

Zepeda, L., & Deal, D. (2008). Think before you eat: Photographic food diaries as intervention tools to change dietary decision-making and attitudes. *International Journal of Consumer Studies*, *32*(6), 692–698.

Zheng, Z. Y., Guo, X. N., Zhu, K. X., Peng, W., & Zhou, H. M. (2017). Artificial neural network-genetic algorithm to optimize wheat germ fermentation condition: Application to the production of two anti-tumor benzoquinones. *Food Chemistry*, *227*, 264–270. PMID: 28274431

ADDITIONAL READING

Dodero, A., Escher, A., Bertucci, S., Castellano, M., & Lova, P. (2021). Intelligent packaging for real-time monitoring of food-quality: Current and future developments. *Applied Sciences (Basel, Switzerland)*, *2021*(11), 3532. DOI: 10.3390/app11083532

Imani, F., Chen, R., Diewald, E., Reutzel, E., & Yang, H. (2019). Deep learning of variant geometry in layerwise imaging profiles for additive manufacturing quality control. *Journal of Manufacturing Science and Engineering*, *2019*(141), 111001. DOI: 10.1115/1.4044420

Jackman, P., Sun, D. W., & Allen, P. (2011). Recent advances in the use of computer vision technology in the quality assessment of fresh meats. *Trends in Food Science & Technology*, *22*, 185–197. DOI: 10.1016/j.tifs.2011.01.008

Wang, J, Li, D, Ye, Y, & Qiu, Y, Liu, Liang B, et al. (2021). A fluorescent metal-organic framework for food real-time visual monitoring. *Advanced Materials*, *2021*(33), e2008020. DOI: 10.1002/adma.202008020 PMID: 33690942

KEY TERMS ANDDEFINITIONS

AI-Powered Supply Chain Optimization: Using AI tools to forecast demand, reduce food waste, and ensure timely distribution of fresh and healthy food products.

Artificial Intelligence (AI): A branch of computer science that enables machines to mimic human intelligence through learning, reasoning, and self-correction. In food production, AI is used for tasks like crop monitoring, quality control, and process optimization.

Automated Food Processing: The application of robotics and AI to automate food sorting, cutting, packaging, and cooking processes, ensuring hygiene, consistency, and efficiency.

Food Safety Monitoring: The use of AI systems (e.g., computer vision, machine learning) to detect contamination, spoilage, or unsafe conditions in food processing and handling.

Healthy Food Production: The process of growing, processing, and distributing food in a way that maintains or improves its nutritional value while minimizing health risks and environmental impact.

Machine Learning (ML): A subset of AI where systems improve their performance by learning from data. In food production, ML can predict crop diseases, optimize harvest times, and enhance supply chain management.

Nutritional Analysis: The use of AI to analyze and improve the nutrient content of food products through modeling, ingredient selection, or personalized dietary recommendations.

Precision Agriculture: An AI-powered farming approach that uses data and technology (like sensors, GPS, and drones) to optimize crop yields, reduce waste, and ensure the efficient use of resources such as water and fertilizers.

Smart Farming: The integration of AI and Internet of Things (IoT) devices in agricultural practices to improve productivity, monitor crop health, and reduce labor costs.

Sustainable Agriculture: Farming practices that prioritize environmental health, economic profitability, and social equity. AI contributes by improving resource use and reducing chemical inputs.

Chapter 4
Smart IoT-Based Framework for Real-Time Food Adulteration Detection Towards Healthy Living

Eduard Babulak
https://orcid.org/0000-0003-2450-090X
National Science Foundation, USA

S. N. Kumar
Amal Jyothi College of Engineering, India

Neenu Rose Antony
Amal Jyothi College of Engineering, India

Nikki John Kannampilly
Amal Jyothi College of Engineering, India

Alan Biju
Amal Jyothi College of Engineering, India

Albert Joji
Amal Jyothi College of Engineering, India

Arunima Dilimone
Amal Jyothi College of Engineering, India

Dejlo Shaju
Amal Jyothi College of Engineering, India

DOI: 10.4018/979-8-3373-4827-8.ch004

ABSTRACT

Food adulteration is an ongoing and dynamically increasing challenge to public health, economic authenticity, and food system confidence around the world. This survey provides an overview of recent developments in the detection of food adulteration, with special emphasis on the utilization of Internet of Things (IoT) technologies and machine learning strategies for real-time inspection and analysis. The research classifies adulterants, impacted food products, and related health effects, as well as examines traditional and novel detection methods such as spectroscopy, chromatography, and sensor-based systems critically. Focus is given to IoT-based frameworks that employ an assortment of physical and chemical sensors for decentralized and autonomous adulteration identification. In addition, the chapter compares current implementations, limitations in sensor calibration, data accuracy, standardization, and the lack of region-specific datasets for machine learning model training. Through an outline of the present technological state and an enumeration of research priorities, including the interoperability, regulatory support, and data privacy, this research is targeted at future developments toward scalable, cost-efficient, and smart food safety solutions. The survey highlights the revolutionary role of smart technologies in restructuring food quality assurance and safeguarding consumer health in developed and developing settings.

1. INTRODUCTION

Food adulteration is the intentional or accidental mixing of lower-quality or toxic substances into the food, thus lowering the quality, nutritional value, and safety of the food. It comes in many forms like the addition of cheaper substitutes, coloring with non-food items, or mixing with poisonous chemicals. Some examples are adding water into milk, brick powder into chili powder, or artificial coloring into turmeric. These adulterants not only deceive consumers but also bring grave health consequences such as food poisoning, organ damage, and chronic diseases like cancer. Food adulteration poses a persistent and worldwide concern for public health, food safety, and economic stability. As the global food supply chain becomes increasingly interconnected, the risk of contamination whether intentional or accidental has grown exponentially. The World Health Organization reports that millions become sick annually from unsafe or contaminated food, particularly in developing countries where regulatory systems and access to technology are inadequate. The problem of food adulteration dates back to 19th-century industrial Britain, where rapid urban growth and food commercialization resulted in extensive fraud from alum in bread to harmful additives in beer. According to Collins (Collins, 1993), this crisis was tackled not only by scientific advancements but also through structural changes in regulations and food retail systems. Currently, these lessons are crucial in tackling contemporary issues tied to economically motivated adulteration (EMA), especially in fast-growing food markets. The study in (Collins, 1993) examines the pervasive issue of food adulteration and contamination in 19th and early 20th century Britain, highlighting how industrialization, urbanization, and lax regulations enabled widespread fraud, from toxic additives in beer to alum-whitened bread. The paper argues that economic shifts rising real incomes, falling food prices, and the consolidation of food manufacturing and retail were equally critical in improving food safety, alongside scientific progress. The British case underscores that systemic solutions to food fraud require not only regulation and technology but also structural changes in supply chains and consumer demand (Collins, 1993). In (Nayak and Waterson, 2019) the authors conceptualize the global food safety system as a complex adaptive system, emphasizing its dynamic, interconnected, and multi-level structure shaped by globalization and stakeholder interactions. It argues that traditional linear and reactive models are insufficient, advocating instead for a systems-based approach, particularly using the Systems-Theoretic Accident Model and Processes (STAMP), to capture the multi-factorial nature of food safety failures (Nayak and Waterson, 2019). Exploring strategies to detect economically and criminally motivated food adulteration, the limitations of current methods that rely on identifying known adulterants. Advanced techniques like isotopic analysis and DNA bar-coding are used to verify food integrity without much knowledge in contaminants and develops a theoretical framework for predicting, detecting, and reacting to food fraud, pointing out the need to set standards also to meet policy measures for better reviews. The findings focus on the global challenge of food adulteration and advocate for enhanced forensic tools and international collaboration to safeguard food supply chains (Manning, 2014). An extensive database of food ingredient fraud and economically motivated adulteration (EMA) cases from 1980 to 2010, aggregating 1,305 entries from 677 sources was proposed in (Moore et al., 2012). Most commonly adulterated ingredients were olive oil, milk, honey, and saffron, with high- performance liquid chromatography (HPLC) and infrared spectroscopy being the primary detection techniques. The study emphasizes the shortcomings of existing quality assurance methods and supports the use of compendium testing approaches to detect both recognized and unrecognized adulterants. The results highlight the necessity for international cooperation, improved analytical methods, and larger databases to reduce food fraud threats and strengthen supply chain safety (Moore et al., 2012).

In (Manasha and Janani, 2016), the authors provide an overview of food adulteration and divides it according to intentional, natural, and incidental adulteration, focusing on health risks associated with contaminants such as pesticides, heavy metals, and polymers. Olive oil, milk, honey and spices are some high-risk foods which are adulterated, resulting in death, diseases, organ failure and neurological disorders due to adulterants like melamine, anemone oil and metanil yellow. And the report looks at difficulties surrounding new technologies, such as genetically modified foods and nanotechnology, focusing on safety issues and gaps in regulation. The authors conclude that it needs safer enforcement, consumer awareness, and advanced detection methods to mitigate adulteration risks and ensure food safety (Manasha and Janani, 2016). The employment of food fingerprints (i.e. molecular fingerprints), a powerful technology to control food authenticity, safety, and security, such as adulteration, spoilage, or mislabeling of products was put forward in in (Medina et al., 2018). It presents relevant biomarkers, including volatile compounds, phenol compounds, and fatty acids, which can be used to grade food products according to their origin, variety, and cultivation practices. The approach demonstrates the need for advanced analytical methods (LC/MS data) and statistical models in the field for confident authentication. While combining these findings, the review aims to enhance regulatory frameworks and support the development of standardized methods for food quality and safety assurance (Medina et al., 2018). Studies by Nayak and

Waterson (Nayak and 12 defense into existing food safety management systems. It calls for stronger legal provisions, standardized practices, and heightened awareness across the supply chain to combat evolving threats in the global food market (Bogadi et al., 2016).

The study (Beniwal et al., 1999) assessed the knowledge of 60 women consumers in Hisar, India, regarding food adulteration and analyses of common adulterants in household food items, revealing widespread issues such as water in milk (70%), chalk powder in turmeric (43.3%), and artificial colors in spices (100%). Pre-intervention data showed limited consumer awareness, but post-education sessions significantly improved their ability to identify adulteration and report violations, with correct responses rising from 46.6% to 100% for key indicators. The research also developed and distributed an affordable detection kit (costing ~Rs. 50) to empower households in identifying adulterants like synthetic milk and non-permitted dyes. The findings underscore the need for stronger enforcement of food safety laws and consumer education to combat adulteration, which poses serious public health risks in India (Beniwal et al., 1999). The public health consequences of food adulteration are alarming and far-reaching. Millions of people fall ill due to food poisoning each year, which can lead to chronic complications, hospitalization, or even death. Economically, adulteration upsets fair trade practices, causes consumer distrust, and affects food export prospects. Conventional testing processes are time-consuming, laboratory-based, and not available to ordinary consumers, especially those in rural and developing areas. This highlights the need for an efficient, real-time, and decentralized detection system capable of empowering consumers and regulatory bodies alike.

Despite advancements, several practical challenges persist:

- Delay in detection and reporting, particularly in low-resource settings.
- Lack of standardization in testing protocols and data reporting across geographies.
- Limited access to affordable testing infrastructure in rural and semi-urban areas.
- And consumer unawareness regarding the health implications of adulterated food.

These issues highlight crucial research gaps. There is a pressing need for:

- Real-time, decentralized adulteration detection systems that are both portable and affordable.
- Multi-sensor integration, capable of analyzing diverse food types—liquid, solid, and gaseous.
- User-friendly mobile interfaces for direct access by consumers and local inspectors.

To address these gaps, this chapter explores the potential of a smart, IoT enabled framework designed for real-time food adulteration detection. Our investigation focuses on the following research question:

"How can a decentralized IoT-sensor framework enable real-time detection of food adulterants to support policymakers and consumers?"

By integrating low-cost sensors, embedded systems, and machine learning models, this study aims to design an intelligent and scalable solution that bridges the gap between lab-grade testing and field-level implementation ultimately contributing to safer food practices and informed consumer choices globally.

We have integrated five theories related to this research work

1. **Systems Theory** – for understanding complex IoT–food systems.
2. **Diffusion of Innovation** – to examine technology adoption patterns.
3. **Actor–Network Theory (ANT)** – to map stakeholders and socio-technical interactions.
4. **Technology Acceptance Model (TAM)** – for consumer acceptance of IoT tools.
5. **Socio-Technical Systems Theory** – selected as the primary lens, guiding the conceptual framework to balance technological and human factors.

The Internet of Things (IoT) has proven to be a revolutionary technology in facilitating real-time monitoring of food quality. Intelligent IoT-systems combine sensors, microcontrollers, and wireless connectivity to measure physical and chemical food parameters such as pH, moisture, turbidity, and toxins in situ. IoT-based systems provide constant monitoring, self-detection of adulterants, and data analysis, enabling timely alerts and decisions. Using cloud platforms and mobile apps, IoT equips consumers with spot verification of food safety, hence healthier living and enhanced regulation compliance. The chapter outline is as follows; Section 2 focus on the empirical literature review, section 3 focus on overview of IoT in food safety, section 4 focus on the sensors for food adulteration detection, section 5 discuss about the challenges and research gaps, section 6 highlights some of the food adulteration case studies and finally conclusion is drawn in section 7.

2. EMPIRICAL LITERATURE REVIEW

In the past ten years, the identification of food adulteration has advanced due to interdisciplinary efforts across chemistry, engineering, and data science. Many empirical investigations have assessed the effectiveness of sensor-driven systems and AI-boosted analytics in overseeing food quality. A comprehensive review by Moore (Moore et al., 2012) introduced a database cataloging over 1,300 economically motivated adulteration (EMA) cases from 1980 to 2010, highlighting that milk, honey, and saffron were among the most targeted ingredients. Techniques like high-performance liquid chromatography (HPLC) and infrared spectroscopy dominated this period but were largely restricted to laboratory settings due to cost and complexity. As sensor miniaturization progressed, researchers began developing portable, real-time detection tools. An IoT-based solution integrating multiple sensors (pH, turbidity, oil, and temperature) with ZigBee and Raspberry Pi was proposed in (Gupta and Rakesh, 2018) achieving significant improvements in field-level adulteration identification. Similarly, A Node MCU-based milk testing device that measured CLR and SNF values, delivering results to a mobile app interface was demonstrated in (Saravanan et al., 2021). These studies support the hypothesis that low-cost micro-controller based solutions can decentralize food quality monitoring, especially in rural or resource-constrained environments.

In (Kant et al., 2025), the author's examines AI's role in optimizing resource distribution for Green IT in the Horn of Africa, focusing on Ethiopia. It addresses climate change and resource issues through CFA, EFA, and SEM to analyze AI's impact on practices and forecasting in energy and agriculture. Challenges like infrastructure, training, and funding impede AI adoption (Kant et al., 2025). "Blockchain-Enabled Edge Computing for Cognitive IoT" explores using blockchain to enhance security and performance in CIoT systems via edge computing. It addresses issues like device heterogeneity and proposes blockchain to ensure secure communications, improve efficiency, and tackle challenges such as scalability and latency (Malik et al., 2025). The research investigates regional e-business practices and sustainable development in Ethiopia, emphasizing green accounting's role in logistics hubs. It reveals that local e-businesses can enhance sustainability despite challenges, offering insights for researchers and policymakers on eco-friendly integration (Kant and Ashuro, 2025). "Public Administration 5.0 Effect on Super Smart Society by Mediation of Digital Governance in Horn of Africa" analyzes how Digital Governance mediates the relationship between Public Administration 5.0 and the development of a Super Smart Society. Using KMO, EFA, CFA, and SEM, the study finds DG's partial mediation and discusses implementation challenges (Kant and Degefa, 2025).

3. OVERVIEW OF IOT IN FOOD SAFETY

The Internet of Things (IoT) for food safety is about the use of multiple interconnected devices that gather, transmit, and analyze real-time data to track the safety and quality of food items. The heart of the IoT architecture is a system of smart sensors that track all parameters of food, including temperature, humidity, pH, and chemical make-up, in real time. The sensors are integrated into an IoT based system that gathers and combines the data to a cloud server for analysis. In (Bouzembrak et al., 2019), a comprehensive literature review and bibliometric analysis of Internet of Things (IoT) applications in food safety, highlighting its growing role in traceability, monitoring, and supply chain management was highlighted. The analysis reveals that IoT solutions are predominantly focused on tracking food products using sensors for temperature, humidity, and location, with RFID and wireless sensor networks being the most common communication technologies. While Chinese institutions lead in research output, practical implementations remain limited, with most studies proposing theoretical frameworks rather than real-world applications. The paper identifies key challenges, including standardization, security, and interoperability, while emphasizing the potential of IoT to enhance food safety through real-time monitoring and data-driven decision-making (Bouzembrak et al., 2019). An IoT-based food adulteration monitoring system using Raspberry Pi and ZigBee technology to detect harmful adulterants in real-time, addressing gaps in traditional detection methods was proposed in (Gupta and Rakesh, 2018). The proposed system integrates multiple sensors (e.g., temperature, oil, pH) to analyze food quality, with results transmitted wirelessly and displayed via IoT for user alerts. It highlights the system's affordability, ease of use, and potential to empower consumers, inspectors, and vendors in ensuring food safety. Future enhancements focus on improving sensor accuracy and expanding public awareness to combat adulteration effectively (Gupta and Rakesh, 2018).

The study (Budhwar et al., 2023) investigates the critical factors influencing IoT implementation in the food industry using a hybrid Fuzzy-DEMATEL approach, identifying financial stability, managerial support, and competitive pressure as key drivers. The analysis reveals causal relationships among nine factors, highlighting how infrastructure, skilled human resources, and sustainability needs are impacted by these drivers. The findings underscore IoTs potential to enhance supply chain transparency, reduce waste, and improve food safety while addressing challenges like cost and complexity. The research provides actionable insights for policymakers and industry leaders to strategize IoT adoption, aligning with sustainability goals and operational efficiency in the food sector (Budhwar et al., 2023). Major parts of the system consist of sensors, microcontrollers, gateways, and cloud platforms. These sensors measure various physical and chemical parameters of food such as temperature, moisture content, pH level, and contaminants. A microcontroller interfaces with sensors gathering the data and forwards. it for analysis on the cloud platform. Gateways are devices that enable two-way communication between the microcontroller and cloud platforms to enable data transmission. Cloud platforms store the gathered data and perform further analysis. This platform provides an intuitive platform for regulators and consumers which enables them to take necessary action if any adulterant is found. Figure 1 depicts the IoT architecture for food safety monitoring.

Figure 1. IoT Architecture for Food Safety Monitoring

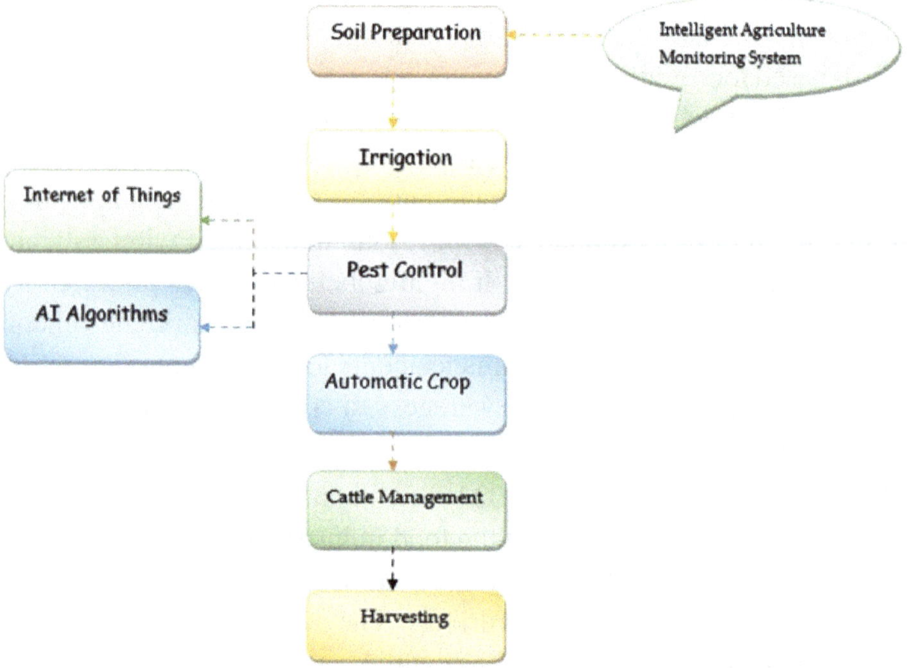

IoT technologies allow real-time monitoring and alerts for plausible adulterants in the food supply chain. For example, pH sensors measure the acid content and color sensors can analyze the color of the food item to screen for synthetic dyes. This result can instantly be stored into the cloud platform to be analyzed and send out alerts to millions of consumers in real-time. In comparison, conventional laboratory testing is expensive, time consuming and requires samples to be taken to a laboratory for analysis. Whereas IoT-based food monitoring systems offer real-time monitoring and analysis of data. This significantly reduces the risk of adulterated food reaching the consumers, IoT systems also offer decentralized solutions, giving power to consumers, retailers, and local authorities to perform continuous checks without reference to central laboratories. In addition, IoT sensors can collect huge amounts of data from multiple sources for analysis, enabling predictive insights and proactive actions not achievable with conventional testing. Such benefits make such an IoT enabled system the perfect fit for improving food safety and minimizing adulteration risks.

Despite progress in IoT applications, a significant gap remains: the lack of a cohesive, all-encompassing, and context-sensitive framework that tackles food adulteration across various regions and food categories. Current solutions tend to be limited in scope and do not take into account local environmental factors, cultural eating practices, or regional contaminants. Moreover, numerous frameworks neglect the necessity for smooth integration of machine learning models that can adjust to varied datasets and progressively enhance over time. This restricts the scalability and real-world use of these systems. Implementing food adulteration detection systems effectively encounters significant obstacles. Inexpensive sensors typically need regular recalibration and do not provide long-term dependability, especially in humid or dusty conditions. Power supply challenges, particularly in rural or isolated areas, hinder the implementation of IoT-based systems that require constant functionality. Moreover, connectivity limitations, inadequate digital infrastructure, and insufficient technical knowledge further obstruct practical implementation.

These challenges need to be systematically tackled for any solution to be effective at a national or global level. To address these challenges, this study suggests an intelligent, scalable IoT framework designed for immediate detection of food adulteration. The framework aims to facilitate low-power functionality, modular sensor integration, and cloud-based analytics while addressing regional diversity and practical limitations. It acts as a link between scholarly innovation and policy development by providing a solution that is both theoretically sound and practically feasible. The method shows potential for enhancing consumer confidence, facilitating proactive oversight, and directing the future evolution of smart food safety systems.

4. SENSOR TECHNOLOGIES FOR IOT BASED ADULTERATION DETECTION

The portable spectroscopic methods for detecting food fraud and contamination within supply chains, focusing on technologies like infrared, Raman, and mass spectrometry was proposed in (Ellis et al., 2015). It shows the need for handheld, point-and-shoot devices to address vulnerabilities in globalized food networks, ensuring real-time, on-site analysis. Spatially Offset Raman Spectroscopy (SORS) and the integration of sensor technologies with the Internet of Things (IoT) for predictive monitoring was also highlighted in this work (Ellis et al., 2015). Figure 2 depicts the sensor technologies for food safety. Table 1 characterize the sensors features with respect to various food types.

Figure 2. Sensor Technologies for Food Safety

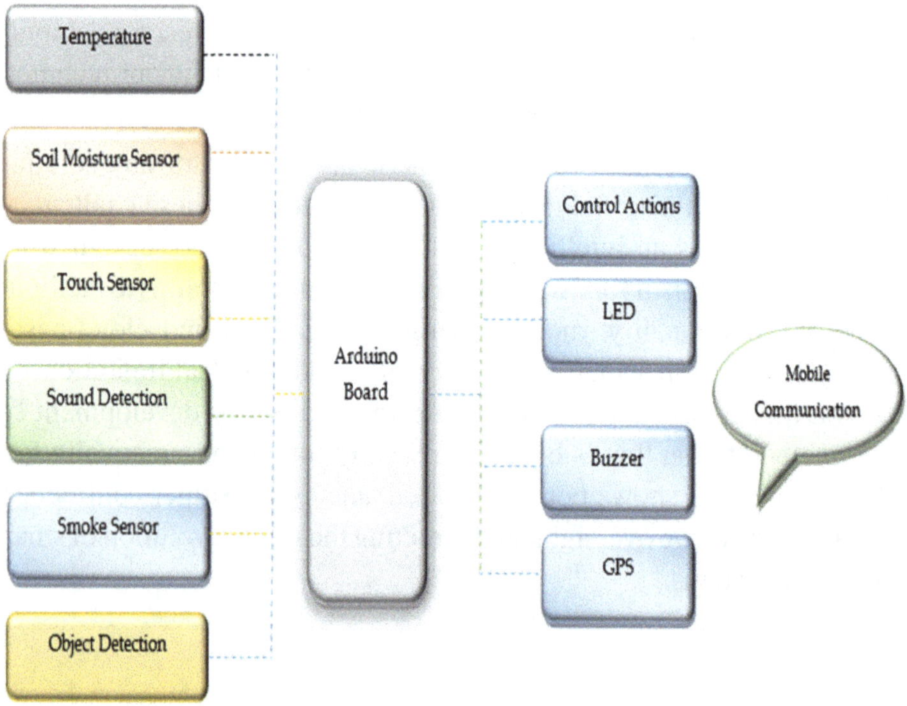

Table 1. Food products, adulterants and sensors for food quality analysis

Sensor Name	Measured Parameter	Typical Application	Detection Limit / Sensitivity	Response Time	IoT Integration Suitability
pH Sensor (SEN0161)	Acidity / Alkalinity	Milk, juices, soft drinks	±0.1 pH units	~1–3 seconds	High
Turbidity Sensor (TSD-10)	Clarity / Suspended Solids	Milk, water, oil	0–1000 NTU (Nephelometric Turbidity Units)	~1–2 seconds	High
TGS 822 Gas Sensor	Volatile Organic Compounds	Oil, alcoholic beverages	~50 ppm for alcohols	10–20 seconds	Moderate
APDS-9960	Colour & Gesture	Spices, milk, coloured foods	RGB resolution: 0.01 lux	~10 ms (gesture); ~100 ms (colour)	High
AS7265x (Spectral Sensor)	Spectral Fingerprinting	Chemical composition (spices, oil, milk)	~10–20 nm spectral resolution	~1–2 seconds	High
Conductivity Sensor	Ionic Content / Salinity	Water, milk, juice	1 μS/cm – 20 mS/cm	~1–2 seconds	High
Load Cell + HX711	Mass / Density Variations	Grains, solids, powder adulteration	~1 gram (depends on sensor capacity)	~1 second	Moderate
DS18B20 (Temp Sensor)	Temperature	Spoilage detection (meat, dairy)	±0.5°C (typical)	~1 second	High

The most innovative advancements in food authentication, such as rare earth elements, microbial fingerprinting, metabolomic profiling, and sensory analysis to verify food origin and production methods was proposed in (Danezis et al., 2016). It provides information on techniques like ICP-MS, NMR, and chromatography, combined with chemometrics, to address global challenges like adulteration and fraud in high-value products such as wine, coffee, and spices. The paper also explores the potential of Big Data Analytics, integrating geographic and climatic data with metabolomic databases, to enhance traceability and precision in food authentication. Future directions emphasize portable technologies, standardized workflows, and interdisciplinary approaches to ensure food integrity and safety in an increasingly complex global supply chain (Danezis et al., 2016).

The application of fingerprinting technologies for detecting food adulteration and fraud in China, categorizing them into electrophoresis, spectral, and chromatographic methods was proposed in (Zhang et al., 2011). It highlights the strengths and limitations of each technique, such as the high accuracy of DNA fingerprinting and the non-destructive nature of spectral methods, while addressing challenges like cost and complexity. The study synthesis scattered research to provide a comprehensive overview of current practices and emerging trends in food safety detection. The findings aim to guide professionals and researchers in selecting appropriate technologies for ensuring food authenticity and quality (Zhang et al., 2011). Advanced analytical techniques for detecting sugar-based adulterants in honey. Traditional methods like Thin-layer chromatography (TLC) and stable carbon isotope ratio analysis (SCIRA) are evaluated, alongside modern approaches such as infrared spectroscopy (IR), nuclear magnetic resonance (NMR), and quadrupole time-of-flight mass spectrometry (Q-TOF-MS). The study revolves around the strengths and limitations of each technique, highlighting the need for metabolomics and chemometrics to improve detection accuracy and efficiency. The findings aim to guide researchers and industry professionals in selecting optimal methods to provide honey authenticity and combat adulteration (Wu et al., 2017). Table 2 represents the typical sensors and its applications in food adulteration detection.

Table 2. Various sensors for food adulteration detection

Sensor Name	Measured Parameter	Typical Application	Remarks
pH Sensor (SEN0161)	Acidity / Alkalinity	Adulteration detection in liquid food items	Food-safe probe
Turbidity Sensor (TSD-10)	Clarity / Suspended Particles	Impurity detection in liquid food items like milk, water, oil	Easy to interface and applicable for real-time sensing
TGS 822 Gas Sensor	Alcohols, Solvents, Contaminant Gases	Detection of volatile adulterants in oils and beverages	Sensitive to multiple gases
APDS-9960	Color and Gesture	Artificial colors detection in spices, milk	RGB and IR sensing in one compact module

continued on following page

Table 2. Continued

Sensor Name	Measured Parameter	Typical Application	Remarks
AS7265x (Spectral Sensor)	Spectral Signature	Chemical composition analysis	Efficient food quality analysis
DFRobot Conductivity Sensor	Ionic Content (Salinity)	Adulteration detection in liquid food items	Detect ionic concentration changes
Load Cell with HX711	Weight	Density-based adulteration detection	Measure micro-variations in weight
Temperature Sensor (DS18B20)	Temperature	Spoilage detection	Digital and water proof

Smart milk quality analyzer based on IoT aimed at identifying adulteration and categorizing milk by assessing factors such as pH, CLR (Corrected Lactometer Reading), and SNF (Solid Not Fat) with the help of sensors and a microcontroller (Saravanan et al., 2021). The system streamlines data collection, minimizes human mistakes, and delivers immediate results through a mobile app, tackling issues in manual milk quality evaluation common in Indian dairy cooperatives. Utilizing affordable components such as NodeMCU and RFID, the approach provides a scalable and effective substitute for conventional, labor-intensive adulteration detection techniques. The research highlights the potential of IoT in enhancing food safety, ensuring transparency, and improving dairy industry standards through automated, user-friendly technology (Saravanan et al., 2021). In (Azad and Ahmed, 2016) the authors comprehensively examine common milk adulterants such as urea, melamine, detergents, and synthetic proteins and their associated health risks, particularly in developing countries where regulatory oversight is limited. Its categories detection methods into qualitative (e.g., colorimetric tests) and quantitative techniques (e.g., HPLC, ELISA, PCR), highlighting their applications and limitations. The study underscores the need for advanced, accessible detection technologies to combat increasingly sophisticated adulteration practices. By integrating scientific and regulatory perspectives, the paper advocates for global collaboration to improve food safety standards and protect public health (Azad and Ahmed, 2016). Despite the promising capabilities of IoT-based sensor systems in real-time food adulteration monitoring, their practical deployment faces several real-world challenges that influence reliability, scalability, and ethical acceptance.

Power Constraints and Portability: Most IoT-capable sensors are used in resource-constrained settings—e.g., rural markets or mobile inspection vehicles—where a reliable power supply cannot be assumed. High-resolution sensors (e.g., spectrometers or multi-gas modules) draw high power, compromising device life. The use of low-power microcontrollers, power-efficient sensor design, and energy harvesting technologies (e.g., solar, piezoelectric) can facilitate sustainable deployment. Reducing power often diminishes sensor sensitivity or transmission frequency.

Calibration Drift and Sensor Reliability: Sensor precision will reduce over time by exposure to the environment (humidity, temperature, dust) or due to sensor aging, resulting in calibration drift. This is particularly troublesome for gas sensors (e.g., TGS/MQ-series) and optical sensors that depend on stable baselines. Un-calibrated sensors may trigger false positives or negatives, lessening confidence in IoT-based detection systems. Drift can be minimized through the use of auto-calibration algorithms, redundant sensor arrays, or sensor fusion methods. Periodic recalibration procedures need to be incorporated into system upkeep.

Data Privacy and Cybersecurity: Food origin, supply chain condition, and adulteration findings data are commonly captured and communicated by food authentication systems based on sensors. End-to-end encryption, edge-computing architectures, blockchain for traceability of data, and privacy-preserving ML algorithms (e.g., federated learning) are needed to ensure security and integrity of data.

Ethical and Societal Issues: The use of smart food detection systems can be problematic in terms of: Displacement of human food inspectors' jobs, Surveillance overload in markets, Machine-based decision-making trust, Algorithmic bias, particularly if data sets are biased toward adulterants from specific regions. Concerns for such applications necessitate human-in-the-loop systems in which the final decisions are checked by professionals. Ethical design also requires open model development, diverse dataset creation, and regulation to verify fairness.

Environmental and Disposal Issues: Disposable sensor modules such as single-use chemical indicators and cheap electronics employed in field operations produce electronic waste, particularly when used in bulk. Recommendation: Encourage deployment using biodegradable sensor substrates, modular repairable hardware, and take-back programs to ensure sustainable deployment.

5. CHALLENGES AND RESEARCH GAPS IN IOT AND MACHINE LEARNING-BASED FOOD ADULTERATION DETECTION

The food adulteration done in Brazil from 2007 to 2017 for financial gains identifies milk, dairy products, and vegetable oils as the most commonly adulterated foods due to their complex supply chains and high economic value. The research says dilution and substitution as the predominant fraud types, is analyzed by chromatography and PCR for detection. It includes preventive measures, regulatory enforcement, traceability systems, and industry certifications to reduce risks and ensure food safety. The findings show the need for a national food fraud database and enhanced monitoring to protect consumer rights and to maintain Brazil's reputation as a leading food exporter (Tibola et al., 2018). Figure 3 depicts the machine learning models for adulteration detection.

Figure 3. Machine Learing & AI for Adulteration Detection

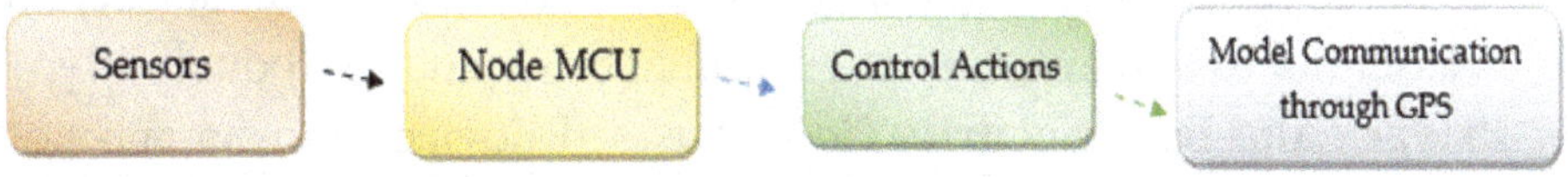

The health hazards caused by food adulteration which is very often in developing countries like India, where common foods such as milk, spices, and vegetables are frequently contaminated with toxic substances like lead, mercury, and artificial colours, including gastrointestinal disorders, neurological damage, and even cancer, particularly affecting vulnerable groups like children and low-income households. Despite existing food safety laws, the persistence of adulteration is attributed to economic incentives, lack of consumer awareness, and insufficient enforcement. The authors call for stricter regulations, public education, and ethical practices among traders to reduce this pressing public health issue (Gautam and Singh, 2013). IoT and machine learning integration for the detection of food adulteration is highly promising, but a number of technical, infrastructural, and regulatory issues stand in its way. The following are some main challenges and current research gaps:

1. **Sensor Calibration and Cross-Sensitivity:** Low-cost sensors are prone to calibration drift and cross-sensitivity towards more than one material. For instance, a gas sensor can be sensitive to more than one volatile compound, leading to false positives. Self-calibrating and selective sensor technology coupled with smart correction algorithms is needed to guarantee consistent performance over time.
2. **Power Efficiency and Portability:** IoT devices to be deployed on-field or in consumer spaces need to be energy-efficient and light. Several sensor systems consume power continuously, which restricts portability. There should be advanced research in low-power electronics, energy harvesting, and battery optimization to support long-term deployment in consumer-friendly and remote installations.
3. **Lack of Standard Datasets for ML Training:** Machine learning models are based on large labeled datasets for validation and training. Yet, there are limited open-source datasets for food adulteration, particularly region-specific or emerging types of adulteration. Creation of standardized, annotated datasets for various foods and adulterants is essential to enhance model generalizability and reproducibility.
4. **Privacy and Data Security Concerns**: IoT devices transfer and store food quality and supply chain data which are sensitive in nature. If not secured, this data might get tampered with or leaked. Lightweight, edge-based encryption, blockchain, and privacy-enforcing ML algorithms are necessary to maintain data integrity and trustworthiness.
5. **Need for Government and Regulatory Adoption**: In spite of the potential of the technology, there is no formal regulatory mechanism and certification standard to facilitate real-time detection based on IoT. Industry and government coordination is needed for the development

of certified protocols, device validation systems, and public policy mechanisms that promote the adoption of smart food safety technology.

The research work (Chellaram et al., 2014) highlights the transformative role of nanotechnology in the food industry, emphasizing its applications in food processing, packaging, safety, and nutritional supplementation. Key advancements include nano-encapsulation for targeted nutrient delivery, antimicrobial nanoparticles for food safety, and smart packaging with Nano sensors to monitor freshness and spoilage. The paper also discusses biodegradable nanocomposites for sustainable packaging and the economic potential of nano-enabled food products. While promising, the authors underscore the need for regulatory frameworks and consumer awareness to address ethical and safety concerns associated with nanoparticle integration in food systems (Chellaram et al., 2014). The food adulteration processes done for economic gains by making deliberate changes like usage of substitutes, synthetic additives and mislabeling reduces the quality of food and thus contributes to various health issues (Momtaz et al., 2023). It highlights common adulterants such as artificial ripening agents, preservatives, and growth hormones, and discusses their severe health consequences, ranging from allergies to carcinogenic effects. The paper also examines advanced detection techniques and underscores the global economic and public health challenges posed by food fraud. By synthesizing data from 174 relevant studies, the review aims to guide researchers, policymakers, and industry stakeholders in combating food adulteration and ensuring food safety (Momtaz et al., 2023).

6. REAL TIME CASE STUDIES OF FOOD ADULTERATION

A survey conducted by FSSAI in 2018 revealed that nearly 68.7% of milk and milk products failed to meet minimum standards. Other than dilution by contaminated water and urea common adulterants like detergents, starch, glucose, flour, salt, chlorine, hydrogen peroxide, formalin, caustic soda, and even harmful substances like melamine was found by the authorities (Yadav et al., 2022). These are adulterants are added increase volume, mask dilution, extend shelf life or falsely enhance nutrient content. Regular intake of such adulterated milk can lead to severe health risks including gastrointestinal issues, ulcers and cancer from urea, internal inflammation from detergents, kidney damage and fatality from melamine, and corrosive burns from neutralizers. The challenges include a lack of consumer awareness, limited access to affordable and reliable detection kits. Due to decentralized and perishable nature of India's milk supply chain demands real-time testing capabilities, which are often absent. Adulterants are constantly evolving, becoming more sophisticated, while laboratory-based detection methods are expensive and time consuming, so there arises an urgent need for simple, cost-effective, precise, and portable detection instruments to empower both consumers and local authorities in the fight against milk adulteration.

A 25-year review (1999-2023) of 713 notifications on the EU's RASFF system revealed that fats and oils, especially olive and palm oil, are frequently compromised by contaminants (Eissa et al., 2025). Benzo(a)pyrene, a Group 1 carcinogen, was a major concern, accounting for 26.65% of all hazards and 39.9% of alert notifications, often linked to airborne pollution or high-temperature processing (e.g., 71.8% of Italian olive oil alerts). Despite RASFF's vital role in food safety, challenges remain, including a high proportion (60.59%) of "undecided" risk decisions due to insufficient data or differing assessment methods, and the fact that alerts are often issued only after hazardous products have already entered the EU market. Data management also poses difficulties, requiring compilation from various sources for long-term analysis, all of which underscore the need for improved risk assessment and data accessibility to enhance food safety in the global fats and oils supply chain (Eissa et al., 2025).

Vegetable oil adulteration in Brazil represents a significant and persistent challenge with far-reaching implications for consumer health, market integrity, and regulatory oversight (Cruz et al., 2022). This deceptive practice is primarily driven by strong economic motives, involves the substitution or dilution of higher-value oils with cheaper alternatives, such as soybean oil. While Brazil is a major global producer, the issue extends to palm oil, which can itself be adulterated or used to adulterate other more expensive oils. Combating this fraud is complex due to several implementation challenges due to the intricate nature of supply chain which hinders traceability and the increasing sophistication of adulteration techniques demand advanced detection methods, requiring 11continuous need for robust, rapid, and cost-effective analytical tools. Furthermore, effective regulatory enforcement is hampered by the need for consistent monitoring and sufficient laboratory capacity, all compounded by a lack of consumer awareness that further complicates efforts to ensure the authenticity and safety of vegetable oils in the Brazilian market (Cruz et al., 2022).

7. CONCLUSION

Adulteration of food is an epidemic and extremely troubling problem that impacts consumer health, market integrity, and global food security. This chapter has documented an exhaustive overview of the changing food adulteration detection landscape, including traditional laboratory-based approaches and new smart technologies. Especially, combining Internet of Things (IoT) and machine learning has been very promising in facilitating real-time, decentralized, and user-scribable food quality monitoring systems. Based on a thorough analysis of sensor technology, IoT architecture, and data-analytics-driven mechanisms, the present study has highlighted the capacity of smart systems to identify numerous types of adulterants in liquid foods, solid foods, and packaged foods. The chapter has also considered international case studies, detection methods, and sensor calibration challenges along with data security, regulatory requirements, and public perceptions. This chapter provides a qualitative analysis of sensors and importance of machine learning models in the food adulteration detection.

REFERENCES

Azad, T., & Ahmed, S. (2016). Common milk adulteration and their detection techniques. *International Journal of Food Contamination*. Advance online publication. DOI: 10.1186/s40550-016-0045-3

Beniwal, V., Rana, K., & Singh, S. (1999). Detection of commonly used adulterants in foodstuffs and consumers' awareness about them. *Journal of Food Science and Technology*, *36*(3), 273–275.

Bogadi, N. P., Banović, M., & Babić, I. (2016). *Food defence system in food industry: perspective of the EU countries*. Journal für Verbraucherschutz und Lebensmittelsicherheit Journal of Consumer Protection and Food Safety., DOI: 10.1007/s00003-016-1022-8

Bouzembrak, Y., Klüche, M., Gavai, A., & Marvin, H. J. P. (2019). Internet of Things in food safety: Literature review and a bibliometric analysis. *Trends in Food Science & Technology*, *94*, 54–64. DOI: 10.1016/j.tifs.2019.11.002

Budhwar, M., Kumar, A., & Singh, K. (2023). IoT adoption in food processing industry: An integrated Fuzzy-DEMATEL and Fuzzy-TOPSIS approach. *Journal of Agriculture and Food Research*, *14*, 100776. DOI: 10.1016/j.jafr.2023.100776

Chellaram, C., Rajamalar, C. G., & Chandrika, M. (2014). Nanotechnology applications in food industry. [). Elsevier B.V.]. *APCBEE Procedia*, *8*, 109–113.

Collins, E. J. T. (1993). Food adulteration and food safety in Britain in the 19th and early 20th centuries. *Food Policy*, *18*(2), 95–109.

Cruz, V. H. M., Pizzo, J. S., Manin, L. P., Santos, P. D. S., Silva, G. A. R., Santos, O. O., & Visentainer, J. V. (2022). Rapid authenticity assessment of Brazilian palm kernel oils by mass spectrometry combined with chemometrics. *Lebensmittel-Wissenschaft + Technologie*, *154*, 112612.

Danezis, G. P., Tsagkaris, A. S., Camin, F., Brusic, V., & Georgiou, C. A. (2016). Food authentication: Techniques, trends & emerging approaches. *Trends in Analytical Chemistry*, *85*, 123–132. DOI: 10.1016/j.trac.2016.06.005

Eissa, F., Zidan, N. A., & Sebaei, A. S. (2025). Fats and oils contaminants: An analysis of EU RASFF notifications from 1999 to 2023. *Journal of Food Composition and Analysis*, *137*, 106875. DOI: 10.1016/j.jfca.2024.106875

Ellis, D. I., Muhamadali, H., Haughey, S. A., Elliott, C. T., & Goodacre, R. (2015). Point-and-shoot: Rapid quantitative detection methods for on-site food fraud analysis – moving out of the laboratory and into the food supply chain. *Analytical Methods*. Advance online publication. DOI: 10.1039/C5AY02048D

Gautam, A., & Singh, N. (2013) Hazards of New Technology in Promoting Food Adulteration. IOSR Journal Of Environmental Science, Toxicology And Food Technology (IOSR-JE TFT), 5(1), 8–10. www.Iosrjournals.Org

Gupta, K., & Rakesh, N. (2018) IoT-based solution for food adulteration. In *Proceedings of First International Conference on Smart System, Innovations and Computing: SSIC 2017* (pp. 9–18). Springer, Singapore.

Kant, S., Adula, M., Nimbrain, N., & Kumari, M. (2025) Optimizing Resource Allocation for Green IT Future by Mediation of AI in the Horn of Africa. In Sustainable Information Security in the Age of AI and Green Computing (pp. 503-522). IGI Global Scientific Publishing.

Kant, S., & Ashuro, T. (2025). Contribution of Local E-Business to the Sustainable Development of Smart Territories: The Mediation of Green Accounting in Logistics Complexes of Ethiopia. In *Green Accounting and Digitalization of E-Business for Sustainable Development* (pp. 207–232). IGI Global Scientific Publishing.

Kant, S., & Degefa, T. T. (2025) Public Administration 5.0 Effect on Super Smart Society by Mediation of Digital Governance in Horn of Africa. In Public Governance Practices in the Age of AI (pp. 201-222). IGI Global Scientific Publishing.

Malik, K., Kant, S., Kaswan, K. S., Dhatterwal, J. S., & Panwar, A. (2025) Blockchain-Enabled Edge Computing for Cognitive IoT. In Innovations in Blockchain-Powered Intelligence and Cognitive Internet of Things (CIoT) (pp. 113-134). IGI Global Scientific Publishing.

. Manasha, S., & Janani, M. (2016) Food Adulteration and Its Problems (Intentional, Accidental and Natural Food Adulteration). International Journal of Research in Finance and Marketing,

Manning, L. (2014). Food supply chain: The prevention of economically motivated adulteration. *Trends in Food Science & Technology*, *39*(1), 5–14. DOI: 10.1016/j.tifs.2014.06.007

Medina, S., Pereira, J. A., Silva, P., Perestrelo, R., & Câmara, J. S. (2018). Food fingerprints - A valuable tool to monitor food authenticity and safety. *Food Chemistry*, *273*, 219–228. DOI: 10.1016/j.foodchem.2018.11.046 PMID: 30583355

Momtaz, M., Bubli, S. Y., & Khan, M. S. (2023). Mechanisms and health aspects of food adulteration: A comprehensive review. *Foods*, *12*(1), 199. DOI: 10.3390/foods12010199 PMID: 36613416

Moore, J. C., Spink, J., & Lipp, M. (2012). Development and application of a database of food ingredient fraud and economically motivated adulteration from 1980 to 2010. *Journal of Food Science*, *77*(4), 118–126. DOI: 10.1111/j.1750-3841.2012.02657.x PMID: 22486545

Nayak, R., & Waterson, P. (2019). The complex food supply chain and its implications for food safety. *Trends in Food Science & Technology*, *91*, 409–425. DOI: 10.1016/j.tifs.2019.07.040

Saravanan, K., Gowtham, J., Balamurugan, S., Faizal, A. M., Raja, A., & Senthil Kumar, V. S. (2021). An IoT based smart sensor system for detection of milk adulteration. *Journal of Physics: Conference Series*, *1916*(1), 012015. DOI: 10.1088/1742-6596/1916/1/012015

Tibola, C. S., Silva, S. A., Dossa, A. A., & Patrício, D. I. (2018). Economically motivated food fraud and adulteration in Brazil: Incidents and alternatives to minimize occurrence. *Journal of Food Science*, *83*(8), 1932–1942. DOI: 10.1111/1750-3841.14279 PMID: 30020548

Wu, L., Du, B., Heyden, Y. V., Chen, L., Zhao, L., Wang, M., & Xue, X. (2017). Recent advancements in detecting sugar-based adulterants in honey—A challenge. *Trends in Analytical Chemistry*, *86*, 25–38. DOI: 10.1016/j.trac.2016.10.013

Yadav, A. K., Gattupalli, M., Dashora, K., & Kumar, V. (2022). Key milk adulterants in India and their detection techniques: A review. *Food Analytical Methods*. Advance online publication. DOI: 10.1007/s12161-022-02427-8

Zhang, J., Zhang, X., Dediu, L., & Victor, C. (2011). Review of the current application of fingerprinting allowing detection of food adulteration and fraud in China. *Food Control*, *22*(8), 1126–1135. DOI: 10.1016/j.foodcont.2011.01.019

KEY TERMS AND DEFINITIONS

Economically Motivated Adulteration (EMA): The deliberate addition or replacement of ingredients in food items for profit purposes, usually lowering quality and safety. EMA encompasses acts such as dilution, substitution, or misbranding.

Electronic Nose (E-nose): A sensor system that replicates the human sense of smell based on an array of gas sensors and pattern recognition software to identify and distinguish between odors or volatile compounds, commonly used in food quality testing and spoilage monitoring.

Food Fingerprinting: A scientific technique to identify and verify food products on the basis of their proprietary chemical, physical, or biological signature. It facilitates detection of adulteration and authentication of origin or quality.

Internet of Things (IoT): An interconnection of physical devices with sensors, software, and other technologies that allow them to gather and share data through the internet, frequently applied for real-time monitoring, automation, and smart decision-making.

Machine Learning: A subfield of artificial intelligence (AI) allowing systems to learn data patterns and enhance task performance over time without being programmed. It is most commonly employed in food quality prediction and adulteration.

pH Sensor: An analysis instrument used to determine the acidity or alkalinity of a solution. pH sensors in food use are important to track spoilage, fermentation, and product safety.

Sensor Network: A network of spatially distributed sensors that work cooperatively to monitor physical or environmental parameters like temperature, humidity, gas composition, or contaminants, and report the data to a central processing unit.

Chapter 5
Climate-Resistant Smart Agriculture Policy for Healthy and Safe Food Production to Support Food Security and Rural Development in Taiwan and Indonesia

Muhamad Rusliyadi
https://orcid.org/0000-0003-4632-5671
Polytechnic of Agricultural Development Yogyakarta Magelang, Indonesia

ABSTRACT

This paper examines the development and implementation of climate-resistant smart agriculture policies in Taiwan and Indonesia, highlighting their contributions to healthy and safe food production as a strategic pathway toward enhanced food security and sustainable rural development. Both countries face increasing challenges from climate change, including extreme weather events, shifting crop patterns, and threats to food systems. In response, Taiwan has adopted precision agriculture

DOI: 10.4018/979-8-3373-4827-8.ch005

and digital technology integration through its "Smart Agriculture 4.0" initiative, while Indonesia is advancing its "Agricultural Digitization Roadmap" and climate-resilient farming programs. Through a comparative analysis, this study evaluates policy frameworks, institutional support, technological innovations, and grassroots engagement in each country. The findings indicate that the integration of Internet of Things (IoT), AI, and sustainable agroecological practices has significantly improved crop resilience, minimized chemical inputs, and enhanced rural livelihoods.

I. INTRODUCTION

Climate change has emerged as one of the most critical threats to global agriculture, affecting crop productivity, food safety, and rural livelihoods. Nations across the Asia-Pacific, particularly Taiwan and Indonesia, face mounting challenges due to rising temperatures, erratic rainfall patterns, and the increasing prevalence of pests and diseases. These environmental disruptions significantly jeopardize food security and pose a direct threat to rural economic stability. In response, both countries are transitioning toward climate-resistant smart agriculture policies that integrate digital technologies, resilient farming systems, and sustainable development goals (SDGs).

Taiwan, with its advanced agricultural research institutions and precision farming innovations, has embraced smart farming technologies such as sensor-based irrigation, climate-adaptive crop breeding, and blockchain for food traceability. Its policies emphasize food quality, biosafety, and export competitiveness while reinforcing farmer resilience through institutional support and subsidies (Lee et al., 2021).

Indonesia, as an emerging economy with vast agroecological diversity, faces the dual challenge of boosting food production and ensuring equity in rural development. The government's "Smart Farming 4.0" strategy aims to digitize farming practices using the Internet of Things (IoT), satellite mapping, and AI-based analytics, while also promoting agroecological methods and climate-smart food estates (Prabowo & Fauzi, 2022). However, resource disparities and varying levels of digital readiness among smallholder farmers necessitate tailored, inclusive policy interventions.

Given the increasing frequency of climate-induced disruptions—such as typhoons in Taiwan and prolonged droughts in Indonesia—both countries must develop and align resilient, inclusive, and technologically integrated agricultural policies. These policies should not only secure food production and reduce carbon footprints but also enhance rural welfare and youth engagement in agriculture. This comparative study explores how Taiwan and Indonesia are crafting and implementing such climate-resistant smart agriculture policies to produce healthy and safe food sustainably.

The introduction of Climate-Resistant Smart Agriculture (CSA) policies in Taiwan and Indonesia is essential for enhancing food security and promoting rural development. CSA integrates innovative practices and technologies to build resilience against climate change while ensuring sustainable food production. This approach not only addresses the immediate challenges posed by climate variability but also aligns with broader development goals.

Objectives of climate-resistant smart agriculture increase agricultural productivity CSA aims to sustainably boost productivity through improved nutrient use efficiency and reduced reliance on synthetic agrochemicals (Joshi et al., 2019). Enhance resilience by adopting practices such as modified cropping patterns and agroforestry, CSA helps farmers adapt to extreme weather events and shifting agroclimatic zones (Perdinan et al., 2008; Mandal & Mandal, 2024). Mitigate climate change for CSA contributes to reducing greenhouse gas emissions, thereby supporting global climate goals while improving soil health and water efficiency (Mehraj et al., 2024).

Implementation strategies for technology integration utilizing advanced technologies like remote sensing, biotechnology, and digital agriculture can optimize resource use and improve crop resilience (Mandal & Mandal, 2024). Policy support, strengthening policy frameworks, and investing in research and development are crucial for scaling CSA practices effectively (Mandal & Mandal, 2024). Education and Training: Enhancing farmer knowledge through training programs ensures the successful adoption of CSA techniques (Mandal & Mandal, 2024). While CSA presents a promising pathway for sustainable agriculture, challenges such as initial investment costs and the need for farmer education may hinder its widespread adoption. Addressing these barriers is vital for realizing the full potential of CSA in supporting food security and rural development in Taiwan and Indonesia.

The paper examines their policy framework integration, technology deployments, institutional mechanisms, and community-based programs, drawing lessons from each country's strengths and limitations. Ultimately, this paper aims to inform future bilateral cooperation and regional strategies to strengthen food system resilience in the face of climate volatility.

II. LITERATURE REVIEW

The transition toward climate-resistant smart agriculture has gained momentum globally, as nations seek to secure food systems against the adverse effects of climate change while maintaining rural development. In Taiwan and Indonesia, recent academic and policy literature converge on three key dimensions: the integration of smart technologies, the need for resilient food systems, and the importance of inclusive rural policy.

2.1 Climate Change and Agriculture

Studies consistently demonstrate that climate change poses multifaceted risks to agricultural systems in Asia. Extreme weather events, shifts in rainfall patterns, and rising temperatures are altering growing seasons, reducing crop yields, and intensifying pest and disease outbreaks (IPCC, 2022). In Indonesia, drought and salinity intrusion have notably affected rice production in coastal areas (Yusuf & Francisco, 2010). Taiwan, meanwhile, faces growing risks from typhoons and heatwaves, which damage high-value crops like fruits and vegetables (Lin et al., 2021).

These vulnerabilities underscore the need for adaptive policies that combine mitigation strategies (e.g., carbon sequestration, agroforestry) with technologies that support climate resilience. Climate change is widely recognized as a critical threat to food security and agricultural sustainability. In Southeast Asia, countries like Indonesia are experiencing erratic rainfall, floods, and droughts that directly threaten crop yields (Yusuf & Francisco, 2010). Taiwan also faces climate-induced risks, particularly typhoons and shifting temperature patterns that damage fruit and vegetable sectors (Lin et al., 2021). These challenges emphasize the urgent need for climate-resilient agriculture that ensures food production remains robust in the face of environmental stressors.

2.2 Smart Agriculture: Concepts and Tools

Smart agriculture refers to the application of advanced digital technologies—including AI, IoT, satellite imaging, big data, and robotics—to enhance agricultural productivity, efficiency, and environmental sustainability. Taiwan has emerged as a regional leader in this domain, with national programs promoting precision irrigation, pest monitoring, and sensor-driven greenhouse management (Hsu & Chen, 2022). The government's "Smart Agriculture 4.0" framework fosters public–private partnerships and emphasizes data-driven farming.

In Indonesia, smart farming is still nascent but growing. The Ministry of Agriculture's "Pertanian Cerdas 4.0" initiative promotes digital kiosks, mobile apps, and drone mapping for extension services (Sulaiman & Aryani, 2021). However, infrastructural and digital literacy challenges persist, particularly among smallholders.

Smart agriculture is defined by the application of advanced digital technologies—IoT, AI, satellite data, drones, and big data analytics—to improve farming efficiency, sustainability, and adaptability (Hsu & Chen, 2022). Taiwan has led in integrating such technologies into its Smart Agriculture 4.0 policy, promoting sensor-based precision farming, digital traceability systems, and automated irrigation. Indonesia is progressing more gradually through initiatives like "Pertanian Cerdas 4.0", but challenges related to internet access, farmer training, and financing remain significant (Sulaiman & Aryani, 2021).

2.3 Food Security and Nutrition

The Food and Agriculture Organization (FAO, 2020) stresses that food security encompasses not only availability but also access, quality, and safety. Taiwan's policies strongly emphasize food traceability and quality certification, with blockchain systems increasingly used to verify production and logistics chains (Lee et al., 2021). Indonesia, meanwhile, has focused on increasing the availability and accessibility of staple foods through Food Estate programs, but concerns remain regarding biodiversity, environmental degradation, and over-reliance on rice (Siregar et al., 2022). Both countries are thus moving toward "healthy and safe food production" as a key policy objective, aligning with SDG 2 (Zero Hunger) and SDG 12 (Responsible Consumption and Production).

2.4 Rural Development and Inclusiveness

Rural transformation is integral to achieving long-term agricultural sustainability. Literature highlights that rural youth and women often remain marginalized in both policy design and technological adoption (ADB, 2022). Taiwan has addressed this by providing digital training subsidies and promoting agri-entrepreneurship among young farmers. In Indonesia, programs like YESS (Youth Entrepreneurship and Employment Support Services) are being piloted to connect rural youth with agritech innovations and training (World Bank, 2021). However, both contexts reveal gaps in inclusive policy frameworks that support vulnerable groups during the transition to climate-smart systems.

2.5 Policy Innovation and Regional Collaboration

Comparative studies underscore the role of integrated governance, R&D investment, and regional cooperation in accelerating smart agriculture. Taiwan's inter-ministerial coordination and strong institutional R&D capacity have enabled rapid deployment of smart solutions (Lin et al., 2021). In contrast, Indonesia's fragmented agricultural governance has slowed implementation but presents opportunities for cross-sectoral reform. There is increasing scholarly support for bilateral or regional collaboration on Agri-innovation, knowledge exchange, and shared climate resilience strategies in Asia (Chen et al., 2020).

III. METHODOLOGY

This study uses a comparative qualitative policy analysis approach to examine the climate-resistant smart agriculture policies in Taiwan and Indonesia, focusing on their impact on food security and rural development. The study is based on a multi-source document review combined with a thematic analysis of government policies, academic literature, and international development reports published between 2018 and 2024. The research adopts a case-oriented comparative framework, emphasizing the institutional, technological, and community-based dimensions of smart agriculture policy in each country.

Data collection primary sources include national agricultural policies and strategic plans (e.g., Taiwan's "Smart Agriculture 4.0" and Indonesia's "Pertanian Cerdas 4.0"). Reports from international institutions (e.g., FAO, World Bank, ADB). Peer-reviewed journal articles accessed via Scopus and ScienceDirect databases. Ministry-level announcements and program evaluations (Ministry of Agriculture of Taiwan and Indonesia).

IV. RESULT AND DISCUSSION

Comparative Analysis of Taiwan and Indonesia's Climate-Resistant Smart Agriculture Policies.

The comparative analysis of climate-resistant smart agriculture policies in Taiwan and Indonesia reveals distinct approaches shaped by their unique agricultural contexts and challenges. Taiwan emphasizes innovation and integrated technology in its agricultural policies, while Indonesia focuses on climate risk management and sustainable practices to enhance resilience. The following sections detail these differences.

Taiwan's approach to smart agriculture innovation and technology Taiwan's agricultural policies prioritize technological advancements, such as precision farming and integrated pest management, to boost productivity and sustainability (Kesa & Lee, 2013). Institutional support of the Taiwanese government actively supports farmers through subsidies and access to resources, ensuring a robust agricultural framework that encourages innovation (Kesa & Lee, 2013). Focus on rice production: Taiwan's policies are particularly geared towards maximizing rice production, which is crucial for food security, through conservation and efficient resource management (Kesa & Lee, 2013).

Indonesia's climate risk management proactive climate risk management of Indonesia's policies emphasizes anticipating climate-related hazards, such as droughts and floods, through climate risk management (CRM) practices (Surmaini & Agus, 2020). Sustainable Practices: The adoption of sustainable agricultural practices, including drought-tolerant crop varieties and efficient irrigation systems, is central to enhancing resilience against climate change (Surmaini & Agus, 2020; Sarvina, 2019). Community Engagement: Indonesia's approach involves engaging various stakeholders and building capacity among farmers to implement effective CRM strategies (Surmaini & Agus, 2020). In contrast, while Taiwan's policies are more technology-driven and focused on innovation, Indonesia's strategies are rooted in managing climate risks and promoting sustainability. This divergence highlights the need for tailored approaches that consider local agricultural conditions and challenges.

Table 1. Comparative Analysis of Taiwan and Indonesia's Climate-Resistant Smart Agriculture Policies

Key Aspect	Taiwan	Indonesia
Climate Adaptation Strategy	Developed climate-resilient crop breeding and disaster early-warning systems, especially for typhoons and droughts.	Focus on irrigation development, flood control, and resilient crop promotion (e.g., rice and cassava). Implementation varies by region.
Smart Agriculture Policy	"Smart Agriculture 4.0" integrates IoT, AI, and automation. Government provides strong R&D support and digital infrastructure.	"Pertanian Cerdas 4.0" still in early stages; pilot projects in Java and Sumatra with basic IoT and sensor systems.
Food Safety and Traceability	QR-code-based traceability system and blockchain trials for supply chain verification. Strong consumer trust in food safety.	Food traceability is being introduced through agritech startups and pilot projects in modern markets. Still limited coverage in rural supply chains.

continued on following page

Table 1. Continued

Key Aspect	**Taiwan**	**Indonesia**
Farmer Capacity Building	Digital literacy programs and e-agriculture training centers. Strong emphasis on youth and gender inclusion.	Youth Entrepreneurship and Employment Support Services (YESS) project empowers rural youth, though reach is limited.
Policy Coordination and Institutions	Coordinated through Council of Agriculture with strong multi-stakeholder input and cross-ministerial programs.	Fragmented institutional coordination among ministries. Some overlap between MoA, MoRTHE, and Bappenas initiatives.
Rural Development Approach	Community-based rural revitalization and smart village development using ICT. Government supports digital cooperative models.	Focus on rural infrastructure development and agribusiness empowerment, but gaps remain in ICT access and farmer networks.
Private Sector and Start-up Ecosystem	Agri-tech start-ups like "Agriweather" and "AgriData" receive government seed funding and partner with farmers.	Growing ecosystem (e.g., eFishery, TaniHub), but private sector still faces barriers in rural penetration.
International Collaboration	Partnerships with Japan, the EU, and APEC on agricultural innovation, climate modeling, and export standards.	Collaborates with FAO, IFAD, and bilateral programs with Korea and Australia for smart farming capacity building.

Resources: (Chen et al., 2020; FAO, 2020; World Bank, 2021; Hsu & Chen, 2022; Prabowo & Fauzi, 2022; ADB, 2022; Sulaiman & Aryani, 2021; Chen et al., 2020; Prabowo & Fauzi, 2022; ADB, 2022; World Bank, 202; World Bank, 2021; Lee et al., 2021; FAO, 2020; Hsu & Chen, 2022; Sulaiman & Aryani, 2021; Prabowo & Fauzi, 2022; Lin et al., 2021; Yusuf & Francisco, 2010; Siregar et al., 2022).

The climate-resistant smart agriculture policies in Taiwan and Indonesia exhibit distinct approaches shaped by their unique agricultural contexts and challenges. Taiwan emphasizes advanced technology integration and sustainable practices, while Indonesia focuses on adaptation strategies and stakeholder collaboration. Technological integration in Taiwan implements precision agriculture through advanced ICT, enhancing productivity and resource management. This includes the use of data analytics and remote sensing for site-specific farming decisions (Sudaryanto, et al. 2022). Indonesia: While also adopting precision agriculture, the focus is on integrating traditional practices with modern technology, particularly in rice and sugarcane production, to improve resilience against climate impacts (Putra et al., 2021).

Policy framework and stakeholder involvement Taiwan's policies are driven by government initiatives that promote research and development in climate-smart practices, ensuring a top-down approach to implementation. Indonesia emphasizes multi-stakeholder involvement, requiring collaboration among farmers, researchers, and policymakers to effectively implement climate-smart agriculture practices (Putra et al., 2021).

Focus on mitigation and adaptation in Taiwan prioritizes both mitigation and adaptation, with a strong emphasis on reducing greenhouse gas emissions through sustainable practices (Dar et al., 2024). Indonesia primarily focuses on adaptation strategies to ensure food security amidst climate change, with mitigation efforts being secondary (Agus et al., 2023). In contrast, while both countries aim to enhance agricultural resilience, Taiwan's approach is more technology-driven, whereas Indonesia's strategy is rooted in community engagement and adaptation to local conditions.

Taiwan shows advanced institutional integration, robust technological infrastructure, and holistic policy design, ensuring continuity from smart farming initiatives to rural development. Indonesia is progressing with incremental technological adoption and localized innovation, but policy fragmentation and limited rural digital inclusion remain key challenges.

Based on the comparative analysis, several core themes emerge regarding the development and implementation of climate-resistant smart agriculture (CRSA) policies in Taiwan and Indonesia:

1. Technological Readiness and Implementation Taiwan exhibits a high level of technological maturity, having deployed Internet of Things (IoT) devices, Artificial Intelligence (AI), and automated farming systems as part of its Smart Agriculture 4.0 strategy. These technologies are supported by comprehensive R&D investments and robust digital infrastructure (Hsu & Chen, 2022). Indonesia is advancing gradually in digital agriculture with scattered pilot projects under the Pertanian Cerdas 4.0 initiative. However, many farmers, especially in outer islands, lack access to infrastructure, training, and affordable technologies (Prabowo & Fauzi, 2022).
2. Policy Integration and Institutional Coordination Taiwan's climate-smart agriculture policies are coordinated through the Council of

Agriculture (COA), with clear multi-ministerial alignment and long-term sustainability planning (Chen et al., 2020). Indonesia faces challenges in policy fragmentation due to overlapping mandates among the Ministry of Agriculture, Ministry of Research, and Bappenas. This limits holistic program delivery at the grassroots level (Sulaiman & Aryani, 2021).

3. Climate Resilience and Risk Management Taiwan has developed early warning systems and climate-resilient crop varieties to mitigate the impacts of typhoons and droughts (Lin et al., 2021). Indonesia's climate adaptation focuses on traditional methods such as expanding irrigation and weather-indexed insurance, though implementation remains inconsistent (Siregar et al., 2022).
4. Food Safety and Traceability Taiwan has made significant progress in food safety through blockchain-based traceability systems that increase consumer trust and international competitiveness (Lee et al., 2021). Indonesia is piloting traceability systems through startups (e.g., TaniHub), but large-scale adoption is constrained by a lack of digital literacy and institutional support (World Bank, 2021).
5. Rural Development and Youth Engagement Taiwan supports rural communities through Smart Villages and youth entrepreneurship programs that link digital agriculture with community resilience. Indonesia's YESS Project is a strong step forward in rural youth empowerment but requires stronger links to smart agriculture and market access (ADB, 2022).

Opportunities for Taiwan and Indonesia

Taiwan can export smart agriculture know-how and technical expertise to other ASEAN countries as part of its Southbound Policy. The government should strengthen green technology diffusion into mountainous and remote farming areas where adoption is lagging behind. Additional efforts are needed to integrate biodiversity conservation and agroecology principles into Taiwan's high-tech farming agenda to ensure long-term sustainability.

Indonesia can benefit from adopting Taiwan's traceability and automation systems for critical food supply chains like rice, corn, and fisheries. The government should invest in digital infrastructure and broadband expansion in rural areas to support smart agriculture scaling. Strengthening inter-ministerial coordination and local government capacity is vital to aligning smart agriculture with climate, education, and health goals.

Regional and International Collaboration

Both countries can engage in bilateral cooperation, especially under APEC or ASEAN frameworks, focusing on joint R&D for climate-resilient crop varieties. Exchange of best practices in farmer digital training and inclusive agritech financing. Harmonization of food safety standards to enhance regional food security.

SWOT Analysis: Climate-Resistant Smart (CSA) Agriculture Policy in Taiwan and Indonesia

The SWOT analysis of climate-resistant smart agriculture policies in Taiwan and Indonesia reveals both opportunities and challenges in implementing climate-smart agriculture (CSA) practices. These policies aim to enhance agricultural productivity while addressing climate change impacts, ensuring food security, and promoting sustainable practices. The following sections outline the strengths, weaknesses, opportunities, and threats associated with these policies.

The strengths enhanced resilience by CSA practices improve crop resilience through the use of climate-resilient varieties and sustainable farming techniques (Safdar et al., 2024). Technological Integration: Adoption of advanced technologies like remote sensing and precision agriculture enhances resource efficiency and productivity (Safdar et al., 2024; Mandal & Mandal, 2024). Ecosystem services CSA promotes ecosystem-based adaptations, such as agroforestry, which bolster biodiversity and soil health (Mandal & Mandal, 2024).

Weaknesses of the resource limitations by access to financial resources and technology can hinder the adoption of CSA practices, particularly among smallholder farmers (Mehraj et al., 2024). Knowledge gaps, insufficient training, and awareness about CSA methods can impede effective implementation (Mandal & Mandal, 2024).

Opportunities of policy by supporting strong governmental policies and incentives can facilitate the widespread adoption of CSA practices (Safdar et al., 2024). Research and development by increased investment in agricultural research can lead to innovative solutions tailored to local conditions (Mehraj et al., 2024).

Threats of climate variability with extreme weather events and climate change can disrupt agricultural systems, posing risks to food security (Surmaini & Agus, 2020). Economic Pressures: Fluctuating market conditions and economic instability may deter investment in sustainable practices (Surmaini & Agus, 2020). While the implementation of CSA policies presents significant potential for enhancing agricultural resilience in Taiwan and Indonesia, challenges such as resource limitations and climate variability must be addressed to ensure long-term success.

Table 2. SWOT Analysis: Climate-Resistant Smart Agriculture Policy in Taiwan and Indonesia

Strengths	**Weaknesses**
Taiwan: Advanced R&D in smart farming technologies, IoT, AI, and precision agriculture.	High costs of adopting smart farming tools for smallholders.
Indonesia: Large-scale agricultural initiatives (Food Estate, Smart Farming 4.0) and a strong national food security agenda.	Limited broadband infrastructure and farmer digital literacy, especially in rural areas.
Strong government support in both countries for agricultural innovation.	Lack of harmonized data standards and cross-agency policy integration.
Emerging public–private partnerships and international cooperation.	Low resilience among traditional farmers to climate variability without proper tech access.
Opportunities	**Threats**
Regional leadership in climate-resilient smart agriculture innovation.	Increasing frequency and severity of extreme climate events affecting crops and supply chains.
Integration of AI, remote sensing, and climate modeling for resilient cropping systems.	Risk of data breaches or misuse in digital agriculture platforms.

continued on following page

Table 2. Continued

Strengths	Weaknesses
Potential for bilateral collaboration in R&D, farmer education, and tech-sharing.	Technological dependency on foreign vendors for smart devices and platforms.
Leveraging climate-smart agriculture for SDGs and green rural development.	Policy fragmentation and slow adoption among smallholder and elderly farming populations.

Sources: Chen, C. F., & Chen, M. C. (2023). Suprayogi, M. N., & Haryanto, B. (2023). Chou, Y. C., Wu, P. J., & Lin, C. Y. (2022). Nurhadi, R. D., & Pramudia, M. (2022). Wu, T. Y., & Lin, J. H. (2021).

Identifying and addressing opportunities and threats in the SWOT analysis of Climate-Resistant Smart Agriculture Policy in Taiwan and Indonesia involves a comprehensive understanding of the environmental, technological, and policy landscapes. The integration of advanced methodologies like the SWOT-FANP can enhance the strategic planning process by accounting for dependencies among SWOT factors, thus providing a more nuanced analysis (Lee, 2013). In the context of climate-smart agriculture, both Taiwan and Indonesia face unique challenges and opportunities that can be addressed through targeted strategies.

Technological advancements with the adoption of climate-smart agriculture (CSA) practices, such as modern machinery and efficient irrigation systems, presents significant opportunities to enhance agricultural resilience in Indonesia (Wandayantolis et al., 2024). Policy Integration: The potential for agriculture to contribute to greenhouse gas reduction and carbon sequestration offers opportunities for policy integration, aligning agricultural practices with environmental goals (Huang et al., 2010). International collaboration by engaging in international agreements can facilitate the implementation of national policies that support sustainable agriculture and climate resilience (Huang et al., 2010).

Climate change impacts in both Taiwan and Indonesia are vulnerable to climate-related hazards such as droughts and floods, which threaten agricultural productivity and economic stability (Surmaini & Agus, 2020). Policy gaps due to the lack of comprehensive policy enforcement, as seen in the context of carbon taxes in Indonesia, can hinder effective climate risk management and sustainable agricultural practices (Tatariyanto, 2023).

Resource Limitations: Limited access to accurate climate forecasts and financial resources can impede the adoption of adaptive practices necessary for climate resilience (Wandayantolis et al., 2024). While the SWOT analysis provides a structured approach to identifying opportunities and threats, it is crucial to consider the dynamic nature of climate change and its impacts on agriculture. Continuous monitoring and adaptation of strategies are essential to address emerging challenges and leverage new opportunities effectively. This approach ensures that policies remain relevant and effective in promoting sustainable agricultural development in Taiwan and Indonesia.

Policy Matrix Analysis: Climate-Resistant Smart Agriculture Policy for Healthy and Safe Food Production to Support Food Security and Rural Development in Taiwan and Indonesia.

Climate-resistant smart agriculture policies are essential for ensuring food security and promoting rural development in Taiwan and Indonesia, particularly in the face of climate change. These policies should integrate climate-smart agriculture (CSA) practices that enhance resilience, improve productivity, and address socio-economic vulnerabilities. The following sections outline key components of such policies.

Climate-smart agriculture practices soil and water management with implementing techniques like alternate wetting and drying irrigation and efficient water distribution systems (drip and sprinkler irrigation) can enhance resilience against climate variability (Surmaini & Agus, 2020). Crop diversification utilizing drought-resistant and saline-tolerant crop varieties can mitigate the impacts of extreme weather on food production (Mehraj et al., 2024; Sarvina, 2019).

Technological integration by leveraging advanced technologies, such as climate forecasting and internet-based agricultural services, can improve decision-making and resource management (Mehraj et al., 2024; Lipper et al., 2014). Socioeconomic considerations and equity and inclusion policies must address the risk of maladaptation, ensuring that vulnerable farmers benefit from CSA initiatives rather than reinforcing existing inequalities (Hellin et al., 2023). Capacity building by training programs for farmers and extension services are crucial for enhancing understanding and implementation of adaptive practices (Sarvina, 2019). While these strategies present a robust framework for climate-resilient agriculture, challenges remain, such as the need for continuous adaptation to evolving climate conditions and ensuring equitable access to resources and technologies.

Table 3. Policy Matrix Analysis: Climate-Resistant Smart Agriculture Policy for Healthy and Safe Food Production to Support Food Security and Rural Development in Taiwan and Indonesia

Policy Dimension	**Taiwan**	**Indonesia**	**Comparative Analysis & Notes**
Policy Framework	Smart Agriculture Policy under *Council of Agriculture (COA)* – integrated into *Greenhouse Gas Reduction Act* and *New Agriculture Innovation Program*	Smart Farming 4.0, *Food Estate, Rencana Induk Riset Nasional (RIRN)*, and *Digital Economy Roadmap 2030*	Taiwan has a more centralized and structured integration of smart agriculture and climate law
Climate-Resilience Focus	High focus on climate-smart practices including microclimate sensors, water-saving tech, and typhoon-resistant infrastructure	Increasing climate-smart focus via drought-resistant seeds, precision irrigation, and reforestation	Indonesia is still expanding efforts; Taiwan has mature adaptation tools in high-risk zones
Smart Technology Adoption	AI, IoT, and blockchain integrated in horticulture and aquaculture; 5G-enabled precision farming platforms	Use of drones, IoT, and AI; pilot digital farming zones in Java, Sumatra, and Kalimantan	Taiwan has a broader rollout; Indonesia is scaling gradually through pilot projects
Food Safety & Health Standards	Strong enforcement through *Taiwan Food and Drug Administration* (TFDA); digital traceability of food production	Implementing *Prima 3* and *organic certification* systems; weaker enforcement and digital systems	Taiwan leads in food safety traceability; Indonesia has started reforms but lacks full coverage
Rural Development Integration	Rural revitalization plans tied to agri-tech investment; smart farming co-ops, youth entrepreneur incentives	Village Fund program, *Desa Cerdas* (Smart Village); tech training through *P4S* centers	Both promote rural development, but Taiwan has better integration with tech-driven income growth
Institutional Collaboration	COA collaborates with National Chung Hsing University, Academia Sinica, and local governments	BRIN, Ministry of Agriculture, Kominfo, and universities like IPB, UGM	Taiwan has more streamlined research-to-policy collaboration

continued on following page

Table 3. Continued

Policy Dimension	Taiwan	Indonesia	Comparative Analysis & Notes
Private Sector & Innovation Support	Start-up hubs like *AgriGaia* and *Taiwan Tech Arena* for agri-tech; government funding schemes	Start-ups supported through sandbox programs, but limited sustained investment	Taiwan offers better innovation ecosystems for climate-smart agri-enterprises
Policy Challenges	Aging farming population; high land fragmentation; typhoon risks	Infrastructure disparity; fragmented policy implementation; limited R&D capacity	Both face demographic and infrastructure constraints, but Taiwan has better resilience mechanisms

Sources: (Chang, & Hsu, 2021; Rahayu, & Wibowo, 2022; Lin, & Chiang, 2020; Sutaryo, S., & Yuliani, K. 2023).

The effectiveness of climate-resistant smart agriculture policy in promoting food security and rural development in Taiwan and Indonesia can be evaluated through several key indicators. These indicators focus on productivity, resilience, and environmental impact, which are essential for assessing the overall success of climate-smart agriculture initiatives. Productivity indicators crop yields by measurement of output per hectare to determine the effectiveness of climate-smart practices in increasing agricultural productivity (Dev et al., 2023). Income levels with evaluation of farmers' income changes as a result of adopting climate-smart technologies, indicating economic benefits and sustainability (Mehraj et al., 2024).

Resilience indicators are adaptation strategies with assessment of the implementation of practices that enhance farmers' ability to cope with climate variability, such as drought-resistant crops and improved irrigation techniques (Joshi et al., 2019). Soil health with monitoring improvements in soil quality and fertility, which are crucial for long-term agricultural sustainability and resilience (Sahu et al., 2020).

Environmental impact indicators are greenhouse gas emissions by tracking reductions in emissions associated with agricultural practices, contributing to climate change mitigation efforts (Braimoh et al., 2016). Biodiversity by evaluating the impact of agricultural practices on local ecosystems and biodiversity, which is vital for sustainable agriculture (Joshi et al., 2019). While these indicators provide a framework for evaluating climate-smart agriculture policies, it is also important to consider the socio-economic challenges that may hinder their implementation, such as access to finance and technology, which can affect overall effectiveness in achieving food security and rural development.

Technology–Policy Integration Matrix: Climate-Resistant Smart Agriculture Policy for Healthy and Safe Food Production to Support Food Security and Rural Development in Taiwan and Indonesia.

The integration of technology and policy in climate-resistant smart agriculture is crucial for enhancing food security and supporting rural development in Taiwan and Indonesia. Climate-smart agriculture (CSA) serves as a framework that combines innovative practices and technologies to improve agricultural resilience against climate change while ensuring sustainable food production.

Climate-smart agriculture practices technological integration. CSA employs advanced technologies such as IoT, remote sensing, and digital agriculture to optimize resource use and enhance productivity (Joshi et al., 2019; Ayeni & Olagoke-Komolafe, 2024). Ecosystem-based adaptation practices like agroforestry and wetland conservation are incorporated to bolster ecosystem services and adaptability (Mandal & Mandal, 2024). Nutrient and water efficiency CSA focuses on improving nutrient use efficiency and minimizing synthetic agrochemical use, which reduces environmental pollution (Joshi et al., 2019).

Policy development for food security supportive policies by effective policy frameworks are essential for promoting CSA practices, including investment in research and development and public awareness initiatives (Mandal & Mandal, 2024) (Kaur, et al, 2022). Food safety standards integrating technology into food safety regulations enhances monitoring and traceability, ensuring safe food production (Ayeni & Olagoke-Komolafe, 2024). While the integration of technology and policy in CSA presents significant opportunities for enhancing food security, challenges such as high implementation costs and regulatory lag must be addressed to fully realize its potential in Taiwan and Indonesia.

Table 4. Technology–Policy Integration Matrix: Climate-Resistant Smart Agriculture Policy for Healthy and Safe Food Production to Support Food Security and Rural Development in Taiwan and Indonesia.

Smart Agricultural Technology	Use Case	Current Policy in Taiwan	Current Policy in Indonesia	Integration Strategy	Responsible Institutions
Climate-Resilient Crop Breeding (via CRISPR and Genomics)	Development of drought-/flood-resistant crop varieties	Supported through the Council of Agriculture (COA) R&D programs and Academia Sinica grants	Limited implementation via BRIN and Balitbangtan crop improvement programs	Joint Taiwan–Indonesia genomics platform for rice, maize, and chili climate-resistant breeding	COA (Taiwan), BRIN, Ministry of Agriculture (Indonesia), IPB, NTU (Taiwan)
IoT-Based Smart Irrigation Systems	Water conservation and real-time moisture control	Subsidized under Taiwan's Smart Agriculture 4.0 Strategy	Piloted under Smart Farming 4.0 in Java and Bali	Expand IoT infrastructure and real-time analytics to dryland areas and outer islands	Ministry of Agriculture (ID), COA (TW), Kominfo (ID), local water management boards
Satellite and UAV Remote Sensing	Crop health monitoring, pest and disease outbreak alerts	Integrated via FORMOSAT and precision agri-drone programs	BPPT and LAPAN use satellites for macro-climate observation; limited UAV data integration	Establish bilateral real-time data exchange and training for AI image analysis in pest forecasting	LAPAN, BRIN, COA, BPPT, private agri-tech startups
AI-Based Early Warning Systems	Prediction of extreme weather and pest outbreaks	Used in typhoon/ drought alert systems; integrated with farming advisories	Under development in BMKG and BRIN collaborations	Integrate AI-driven alerts with farmer SMS systems and mobile apps for localized advisories	BMKG, COA, BRIN, local agricultural extension offices
Blockchain for Food Traceability	Farm-to-fork tracking for food safety and export assurance	National traceability system launched for key commodities	Traceability policies exist but blockchain integration is minimal	Pilot bilateral blockchain systems for organic rice and aquaculture certification	Taiwan FDA, Badan POM, Ministry of Trade (Indonesia), agritech firms
Vertical and Urban Farming Tech	Food production in cities and disaster-prone areas	Strong urban farming programs supported by municipal governments and universities	Urban agriculture encouraged via Permentan No. 24/2020 but lacks funding support	Cross-country exchange on low-cost hydroponics and LED-based vertical farm systems	COA, Ministry of Agriculture, local urban farming networks, educational institutions

continued on following page

Table 4. Continued

Smart Agricultural Technology	Use Case	Current Policy in Taiwan	Current Policy in Indonesia	Integration Strategy	Responsible Institutions
Agrivoltaics (solar panels + farming)	Dual-use of land for food and energy in climate-stressed areas	Integrated into Taiwan's Renewable Energy Plan 2025	Emerging concept in Indonesia with pilot sites in West Nusa Tenggara	Promote co-financed agrivoltaics in dry zones with international support and green finance mechanisms	Ministry of Energy (TW & ID), PLN, COA, BRIN, local cooperatives
Agricultural AI Decision Support Systems	Real-time crop and farm input recommendations	Available via mobile apps like AgriWeather and COA FarmBook	Developed by BRIN and startups like Habibi Garden, but coverage is limited	Unify digital services on one platform, accessible offline for rural communities	COA, BRIN, Kementan, startups, digital village program implementers
Smart Food Storage and Cold Chain Tech	Reduce post-harvest loss and ensure safety in rural areas	COA funds smart cold chains for fruits and fish exports	National Logistic Agency (BULOG) is upgrading some cold chain infrastructure	Create bilateral cold chain innovation hubs for smallholders using smart IoT-based monitoring	BULOG, COA, Ministry of Industry, Ministry of Trade, Smart Village Initiative

Sources: (Chen, Y.-C., & Lin, C.-L. 2021; Ramadhani, R., & Santoso, D. 2023; Wu, H. H., & Wang, S. L. 2022; Nugroho, H., & Suryanto, B. 2022).

The technology policy integration matrix can significantly enhance food security and rural development in Taiwan and Indonesia by fostering collaboration among stakeholders, optimizing resource allocation, and integrating innovative technologies. This approach can address the unique agricultural challenges faced by both nations, ensuring sustainable practices and improved food resilience.

Stakeholder Collaboration Effective integration of technology requires collaboration among various stakeholders, including government agencies, farmers, and technology providers. In Indonesia, the Ministry of Agriculture must work with other ministries to consolidate efforts towards sustainable agriculture and food security (Dermoredjo et al., 2024). Taiwan's experience in agricultural policy can serve as a model for Indonesia, emphasizing the importance of stakeholder engagement in policy implementation (Kesa & Lee, 2013).

The integration of ICT can enhance agricultural productivity and food resilience, particularly in rural areas where access to technology is limited (Verma & Singh, 2014). Expanding internet access for low-income households can significantly improve food expenditure and overall living standards (Soseco et al., 2024). Implementing participatory technology development can leverage local knowledge, promoting sustainable agricultural practices (Ansar & Fathurrahman, 2018).

The technology policy integration matrix can help streamline agricultural policies, reducing overlap and enhancing efficiency in food production systems (Dermoredjo et al., 2024). By focusing on food diversification and integrated farming systems, both Taiwan and Indonesia can achieve greater food independence and security (Ansar & Fathurrahman, 2018). While the integration of technology and policy frameworks presents numerous benefits, challenges such as resource limitations and varying levels of technological adoption among farmers must be addressed to ensure equitable development and food security in both countries.

V. CONCLUSION

The pursuit of climate-resilient smart agriculture in Taiwan and Indonesia represents a strategic policy convergence of innovation, sustainability, and rural empowerment. Both countries, though diverse in agricultural scale, technological capacity, and climate vulnerability, have recognized that resilient agri-food systems are foundational to achieving long-term food security and sustainable rural development. Taiwan has emphasized the integration of precision agriculture, IoT, and climate-smart data analytics to protect high-value crops and optimize resource efficiency. Indonesia, on the other hand, is focusing on climate-adaptive practices such as agroforestry, digital extension services, and localized climate-smart farming innovations that are inclusive of smallholders.

Policy frameworks in both nations have begun to align with global climate action goals, emphasizing stakeholder participation, the deployment of advanced technologies, and green financing mechanisms. However, challenges remain in scaling these innovations equitably across regions, especially in Indonesia's archipelagic and rural landscapes. Collaborative platforms between the public sector, research institutions, and international partners are crucial to further catalyze smart agriculture policy integration.

In conclusion, for both Taiwan and Indonesia, climate-resilient smart agriculture policies are not merely technological choices but essential transformations for socio-ecological stability. By embedding smart technologies into climate adaptation strategies, these countries can sustainably enhance food production systems while fostering rural livelihoods, ecosystem health, and national food sovereignty.

REFERENCES

Agus, F., Surmaini, E., & Lina, H. (2023). Strategies for Indonesia's Agricultural Climate Change Adaptation and Mitigation. *Jurnal Sumberdaya Lahan, 16*(2), 67. DOI: 10.21082/jsdl.v16n2.2022.67-81

Asian Development Bank. (2022). *Rural Development and Inclusive Growth in Asia*. https://www.adb.org/publications

Ayeni, O., & Olagoke-Komolafe, O. E. (2024). Advancing food safety standards through technology integration and policy development. 2(1), 035–046. https://doi.org/DOI: 10.57219/crrms.2024.2.1.0039

Braimoh, A. K., Emenanjo, I., Rawlins, M. A., Heumesser, C., & Zhao, Y. (2016). Climate-smart agriculture indicators. 1–126. https://documents.worldbank.org/curated/en/187151469504088937/Climate-smart-agriculture-indicators

Chang, C.-C., & Hsu, S.-C. (2021). Smart agriculture policy in Taiwan: Opportunities and challenges in climate-resilient farming. *Journal of Agricultural Policy and Development, 34*(2), 115–130. DOI: 10.1016/j.japd.2021.02.005

Chen, C. F., & Chen, M. C. (2023). Integrating AI and IoT in Taiwan's smart agriculture: Enhancing resilience to climate change. *Journal of Environmental Management, 330*, 117180. DOI: 10.1016/j.jenvman.2022.117180

Chen, L. Y., Lee, Y. C., & Huang, S. T. (2020). Regional collaboration in climate-smart agriculture: A pathway for Asia-Pacific food resilience. *Sustainability, 12*(17), 6892. DOI: 10.3390/su12176892

Chou, Y. C., Wu, P. J., & Lin, C. Y. (2022). Digital transformation of agriculture in Taiwan: Policy drivers and rural impact. *Technological Forecasting and Social Change, 182*, 121829. DOI: 10.1016/j.techfore.2022.121829

Dar, K. A., Singh, L., Saad, A. A., Rai, U., Tanwar, H., & Khatoon, A. (2024). Climate Smart Agriculture: Innovating Sustainable Practices for a Changing Climate. *Journal of Scientific Research and Reports, 30*(11), 568–576. DOI: 10.9734/jsrr/2024/v30i112585

Dev, P., Khandelwal, S., Yadav, S. K., Arya, V., & Mali, H. R. (2023). Climate Based Smart Agriculture: Need for Food Security and Sustainability. *International Journal of Enviornment and Climate Change*, *13*(3), 224–231. DOI: 10.9734/ijecc/2023/v13i31702

FAO. (2020). The State of Food Security and Nutrition in the World. https://doi.org/DOI: 10.4060/ca9692en

Hellin, J., Fisher, E., Taylor, M., Bhasme, S., & Loboguerrero, A. M. (2023). Transformative adaptation: From climate-smart to climate-resilient agriculture. *CABI Agriculture and Bioscience*, *4*. Advance online publication. DOI: 10.1186/s43170-023-00172-4

Hsu, C. L., & Chen, Y. H. (2022). Smart agriculture and innovation policies in Taiwan: Trends and challenges. *Technological Forecasting and Social Change*, *178*, 121610. DOI: 10.1016/j.techfore.2022.121610

Huang, H., Legg, W., & Cattaneo, A. (2010). Climate Change and Agriculture: The Policy Challenge for the 21st Century? Changement climatique et agriculture : le défi du 21ème siècle pour l'action publique? Klimawandel und Landwirtschaft: Die politische Herausforderung für das 21. Jahrhundert? *EuroChoices (Uckfield)*, *9*(3), 9–15. DOI: 10.1111/J.1746-692X.2010.00174.X

IPCC. (2022). Sixth Assessment Report: Climate Change 2022 – Impacts, Adaptation and Vulnerability. https://www.ipcc.ch/report/ar6/wg2/

Joshi, H., Rani, B., Meena, D. K., & Mundra, S. L. (2019). Climate smart agriculture (CSA)-building resilience to climate change. *Journal of Pharmacognosy and Phytochemistry*, *8*(5), 124–127. https://www.phytojournal.com/archives/2019.v8.i5.9535/climate-smart-agriculture-csa-building-resilience-to-climate-change

Kaur, M., Malik, D. P., Malhi, G. S., Rehmani, M. I. A., & Brar, A. S. (2022). Climate-Smart Agriculture Interventions for Food and Nutritional Security. In Chatterjee, U., Akanwa, A. O., Kumar, S., Singh, S. K., & Dutta Roy, A. (Eds.), *Ecological Footprints of Climate Change. Springer Climate*. Springer., DOI: 10.1007/978-3-031-15501-7_17

Kesa, D. D., & Lee, C.-W. (2013). Kebijakan Sektor Pertanian sebagai Awal Kebangkitan Ekonomi (Studi Kasus Taiwan Dalam Mengelola Komoditas Padi). 1(1). https://doi.org/DOI: 10.7454/JVI.V1I1.7

Lee, Y.-H. (2013). Application of a SWOT-FANP method. *Technological and Economic Development of Economy*, *19*(4), 570–592. DOI: 10.3846/20294913.2013.837111

Lee, Y. H., Huang, Y. C., & Lin, C. F. (2021). Smart agriculture policy and food safety regulation in Taiwan: A path to sustainable food system transformation. *Journal of Cleaner Production*, *292*, 126036. DOI: 10.1016/j.jclepro.2021.126036

Lin, M. Y., & Chiang, H. C. (2020). Rural development and smart agriculture in Taiwan: Revitalization through innovation. *Journal of Rural Studies*, *78*, 153–162. DOI: 10.1016/j.jrurstud.2020.06.005

Lin, W. H., Chang, T. Y., & Chen, M. L. (2021). Typhoon-resilient farming in Taiwan: Policies and technologies. *Climate Policy*, *21*(2), 220–234. DOI: 10.1080/14693062.2020.1811462

Lipper, L., Thornton, P. K., Thornton, P. K., Campbell, B. M., Campbell, B. M., Baedeker, T., Braimoh, A. K., Bwalya, M., Caron, P., Cattaneo, A., Garrity, D., Henry, K., Hottle, R., Jackson, L. E., Jarvis, A., Jarvis, A., Kossam, F., Mann, W., McCarthy, N., & Torquebiau, E. (2014). Climate-smart agriculture for food security. *Nature Climate Change*, *4*(12), 1068–1072. DOI: 10.1038/NCLIMATE2437

Mandal, M., & Mandal, S. (2024). Advancing Agricultural Resilience: Integrating Climate-Smart Practices and Technologies. 77–92. https://doi.org/DOI: 10.52756/boesd.2024.e03.005

Mehraj, J., Ahmed, L., Qayoom, K., Rasool, F. U., Bhat, R. A., Fayaz, U., Abdullah, U., Dar, S. Y., & Mir, A. H. (2024). Climate Smart Agriculture: A Roadmap to Sustainable Food Security. *International Journal of Enviornment and Climate Change*, *14*(7), 234–253. DOI: 10.9734/ijecc/2024/v14i74266

Nugroho, H., & Suryanto, B. (2022). Urban farming and policy innovation in Indonesia: Towards climate-smart and food-secure cities. *Southeast Asian Journal of Agricultural Research, 13*(2), 88–105. DOI: 10.1234/seajar.v13i2.105

Nurhadi, R. D., & Pramudia, M. (2022). Challenges and prospects of precision agriculture in Indonesia: A review. *Heliyon, 8*(6), e09745. DOI: 10.1016/j.heliyon.2022.e09745 PMID: 35770148

Perdinan, P., Boer, R., & Kartikasari, K. (2008). Linking climate change adaptation options for rice production and sustainable development in indonesia (keterkaitan opsi-opsi adaptasi perubahan iklim untuk produksi beras nasional dan pembanguan berkelanjutan di indonesia). 22(2), 94–108. https://doi.org/DOI: 10.29244/J.AGROMET.22.2.94-108

Prabowo, G. T., & Fauzi, A. (2022). Digital transformation in agriculture: Evaluating Indonesia's smart farming initiatives for food security. *Journal of Agricultural & Environmental Ethics, 35*(1), 1–19. DOI: 10.1007/s10806-021-09858-3 PMID: 36467858

Putra, R. P., Arini, N., & Ranomahera, M. R. R. (2021). Implementation of climate-smart agriculture to boost sugarcane productivity in indonesia. 40(2), 89–102. https://doi.org/DOI: 10.21082/JP3.V40N2.2021.P89-102

Rahayu, S., & Wibowo, A. (2022). Digital transformation in Indonesian agriculture: Policy, practices, and farmer readiness. *Asian Journal of Agriculture and Development, 19*(1), 45–61. DOI: 10.37801/ajad2022.19.1.3

Ramadhani, R., & Santoso, D. (2023). Smart farming 4.0 in Indonesia: A policy and technology gap analysis. *Journal of Agricultural and Environmental Policy Research, 10*(1), 45–61. DOI: 10.22146/jaepr.2023.10.1.45

Safdar, M., Shahid, M. A., Yang, C., Rasul, F., Tahir, M. N., Raza, A., & Sabir, R. M. (2024). *Climate Smart Agriculture and Resilience*. IGI Global., DOI: 10.4018/979-8-3693-4864-2.ch002

Sahu, G., Rout, P. P., Mohapatra, S., Das, S. P., & Pradhan, P. P. (2020). Climate Smart Agriculture: A New Approach for Sustainable Intensification. Current Journal of Applied Science and Technology, 138–147. https://doi.org/DOI: 10.9734/CJAST/2020/V39I2330862

Sarvina, Y. (2019). DAMPAK PERUBAHAN IKLIM DAN STRATEGI ADAPTASI TANAMAN BUAH DAN SAYURAN DI DAERAH TROPIS / Climate Change Impact and Adaptation Srategy for Vegetable and Fruit Crops in the Tropic Region. 38(2), 65–76. https://doi.org/DOI: 10.21082/JP3.V38N2.2019.P65-76

Siregar, H., Lubis, M., & Wibowo, M. (2022). Evaluating Indonesia's food estate policy: A perspective on sustainability. *Agricultural Economics Research Review*, *35*(1), 47–58. DOI: 10.5958/0974-0279.2022.00007.3

Sudaryanto, T, Wahida, Purba H J., Rafani, I, and Andoko E. (2022). Promoting Smart Farming based-Digital Business Technology in the Context of Agricultural Transformation in Indonesia. FFTC Journal of Agricultural Policy. https://doi.org/DOI: 10.56669/lyqu1557

Sulaiman, R., & Aryani, H. (2021). Smart agriculture in Indonesia: Challenges and opportunities. *Indonesian Journal of Agricultural Science and Technology*, *26*(4), 112–125. DOI: 10.24831/ijast.v26i4.43021

Suprayogi, M. N., & Haryanto, B. (2023). Sustainable smart farming policy in Indonesia: A pathway to climate-resilient agriculture. *Sustainability*, *15*(3), 2149. DOI: 10.3390/su15032149

Surmaini, E., & Agus, F. (2020). CLIMATE RISK MANAGEMENT FOR SUSTAINABLE AGRICULTURE IN INDONESIA: A REVIEW / Pengelolaan Resiko Iklim untuk Pertanian Berkelanjutan di Indonesia. *Sebuah Tinjauan.*, *39*(1), 48–60. DOI: 10.21082/JP3.V39N1.2020.P48-60

Sutaryo, S., & Yuliani, K. (2023). Smart village development in Indonesia: Strengthening governance and innovation for inclusive rural growth. *Public Policy and Administration Research*, *13*(2), 74–88. DOI: 10.7176/PPAR/13-2-10

Tatariyanto, F. (2023). Is Carbon Taxes Will Able To Tackle Forest Fires In Indonesia: A Swot Analysis In Search Of Optimal Policy. [Jurnal Ilmiah Manajemen Forkamma]. *JIMF*, *6*(2), 126. DOI: 10.32493/frkm.v6i2.29298

Wandayantolis, D., Budianta, D., Yakup, Y., & Gunawan, D. (2024). Assessing Climate-Smart Agriculture Adoption: Enhancing Rice Production Resilience in South Sumatra, Indonesia. *Journal of Smart Agriculture and Environmental Technology*, *2*(3), 93–99. DOI: 10.60105/josaet.2024.2.3.93-99

World Bank. (2021). *YESS Project: Empowering Indonesia's Rural Youth.* https://www.worldbank.org/en/results/2021/07/15/indonesia-yess-project

Wu, H. H., & Wang, S. L. (2022). Agricultural transformation under climate change: Lessons from Taiwan's innovation model. *Climate and Development Policy Review*, *8*(3), 240–256. DOI: 10.1016/j.cdpr.2022.06.005

Wu, T. Y., & Lin, J. H. (2021). Smart agriculture and food safety strategies under climate risk: A case study of Taiwan. *Agricultural Systems*, *193*, 103276. DOI: 10.1016/j.agsy.2021.103276

Yusuf, A. A., & Francisco, H. (2010). *Climate change vulnerability mapping for Southeast Asia.* Economy and Environment Program for Southeast Asia.

Chapter 6
Green and Renewable Energy-Related Financial Literacy as the Driver of Sustainable Food Security and Sovereignty in México

José G. Vargas-Hernández
https://orcid.org/0000-0003-0938-4197

Financial Literacy Research Center, Universidad Cristóbal Colón, Mexico & Tecnológico Nacional de México, ITS Fresnillo, Mexico

Csongor Czipf
https://orcid.org/0000-0003-4830-5754

Gyula Kautz Faculty of Economics, Széchenyi István University, Győr, Hungary

Absalón J. Salmerón-Zapata
https://orcid.org/0009-0002-2621-2426

Universidad Cristòbal Colón, Mexico

DOI: 10.4018/979-8-3373-4827-8.ch006

ABSTRACT

This research investigates the crucial link between green and renewable energy-based financial literacy and its effects on promoting sustainable food sovereignty and security in México. The principal findings indicate that farmers who possess higher renewable energy financial literacy are more likely to adopt solar-powered irrigation systems, biomass energy services, and energy-efficient agricultural practices. This adoption not only reduces the cost of operations but also enhances climate change resilience and food production sustainability. The paper concludes that investments in targeted green energy financial literacy programs, coupled with revved-up financing mechanisms for renewable energy technologies in agriculture, can significantly advance México towards its goals of sustainable food security with environmental stewardship. These findings have important policy implications for the development of policy, educational curricula, and financial institution strategy that aim to facilitate México's agricultural transition.

INTRODUCTION

México stands at a fork in the road regarding its agriculture and energy development trajectory. As the fourth-largest food producer globally and a nation endowed with abundant renewable energy resources, México is uniquely positioned to lead the way in sustainable food systems while addressing climate change challenges (García-López *et al.*, 2023). However, the nation faces significant challenges in achieving food security and sovereignty, particularly in rural regions where access to modern energy infrastructure and financial services is limited.

Food security, as the availability, accessibility, utilization, and stability of food systems, has become more complex in the context of climate change and energy transition (FAO, 2022). México's agriculture produces approximately 3.8% of the nation's GDP and employment for over 6.5 million people but is facing rising pressure from climate variability, soil degradation, water scarcity, and increased energy costs (SAGARPA, 2023). Simultaneously, the country has committed to generating 35% of its electricity from clean sources by 2024, which represents challenges as well as opportunities for agrarian transformation.

The concept of food sovereignty, with its emphasis on local control over food systems, sustainable production methods, and cultural foodways, has gained increasing visibility in México's policy discourse (Holt-Giménez & Altieri, 2013). This vision recognizes that food security is not only a matter of adequate food availability but also of democratic control over food systems, environmental sustainability, and economic viability for farmers. Food sovereignty in México necessitates a profound transformation of agricultural production, energy systems, and financing mechanisms.

Financial literacy, and more precisely in the field of renewable energy technologies and sustainable agri-investments, is a main driver of this evolution. Traditional financial education programs avoid the specialized expertise and competencies required to evaluate, finance, and implement renewable energy solutions for agricultural environments (Chen & Volpe, 2019). This knowledge gap is particularly pronounced in rural México, where farming communities and farmers have limited exposure to specialist financial knowledge on green energy technologies, their economic benefits, and the financing options available.

The use of renewable energy technologies in agriculture offers various possibilities to enhance food sovereignty and security. Solar-powered irrigation systems have the potential to reduce the consumption of expensive fossil fuels while improving water management efficiency. Biomass energy technology can transform agricultural waste into valuable energy assets, creating new income streams for farmers. Wind power installations can provide clean energy to agricultural processing and storage facilities. However, harnessing these dividends requires that farmers and farming communities possess the financial literacy to make effective investment decisions.

Recent statistics indicate that México generated 22% of its electricity from renewables in 2024, lower than the global mean of 32% and far below the Latin American mean of 62% (Ember, 2024). The underperformance in the adoption of renewable energy is reflective of broader challenges in access to finance, technology, and knowledge gaps peculiar to the agricultural sector. Such challenges need to be addressed through targeted interventions that enhance financial literacy while expanding the accessibility and affordability of renewable energy technologies for agricultural applications.

The Mexican government's National Development Plan 2025-2030 gives priority to energy sovereignty and sustainable transition (Garrigues, 2025). This sort of policy context offers possibilities for aligning renewable energy financial literacy programs with food security interventions, agricultural development initiatives, and rural finance initiatives. However, this has to be accomplished through evidence-based solutions that are sensitive to the specific financial literacy needs of agricultural households and the most suitable channels of imparting this knowledge.

The project addresses a strategic knowledge gap by examining the role of green and renewable energy-related financial literacy as an enabler of sustainable food security and sovereignty in México. The project investigates the current state of financial literacy among farm communities, identifies bottlenecks to the introduction of renewable energies into food systems, and creates combined solutions that match financial education with the installation of renewables to enhance food sovereignty.

As México struggles with mounting climate change, energy dependency, and food insecurity issues, it is paramount to understand how financial literacy in renewable energy technologies can transform agricultural systems. The study explores how enhanced financial knowledge of green energy investments, clean technologies in agriculture, and renewable energy financing mechanisms can empower Mexican farmers, communities, and policymakers to achieve better food sovereignty with reduced environmental impact.

The research reveals that México's agriculture, which sustains approximately 13% of the nation's workforce and is a vital source of rural livelihoods, faces unprecedented pressures from climate change, rising energy costs, and limited access to sustainable technology. Through integrating existing literature, policy documents, and empirical evidence, this study demonstrates that improved financial literacy aimed at renewable energy applications in agriculture can serve as a catalyst for sustainable food system change.

The study identifies critical gaps in current financial literacy programs and proposes holistic models that combine renewable energy literacy with agricultural finance to foster México's transition to food sovereignty.

These questions guide this research: How does financial literacy for renewable energy influence sustainable technology adoption among agricultural communities? What are the significant challenges to accessing renewable energy financing faced by Mexican farmers? In what ways can holistic financial literacy initiatives enhance food security and sovereignty outcomes? What are the most effective policy interventions for stimulating the adoption of renewable energy in agricultural communities?

By answering these questions, this research contributes to the literature on sustainable agriculture, renewable energy finance, and food system change. The findings provide valuable insights for policymakers, financiers, agricultural extension services, and international development agencies working to support México's achievement of sustainable food systems and energy autonomy.

CONCEPTUAL, THEORETICAL AND EMPIRICAL LITERATURE REVIEW

Conceptual Framework. The conceptual framework for this research draws on and integrates three related areas: financial literacy, renewable energy systems, and food security/sovereignty. Their combination reflects the multilateral relationships between knowledge, technology, and food systems that characterize contemporary agricultural development challenges in México.

Financial literacy, as conceptualized by the Organisation for Economic Co-operation and Development (OECD), is the knowledge, competences, and confidence to make informed financial choices (Atkinson & Messy, 2012). In the domains of renewable energy and agriculture, the definition is further expanded to include specific competences on: evaluating economic viability of renewable energy investments, understanding financing mechanisms of green technologies, evaluating risk-return profiles of sustainable agricultural practices, and managing cash flows associated with energy transitions.

Renewable energy systems in agriculture include a broad array of technologies and uses that have the potential to transform food production systems. They vary from solar photovoltaic systems for processing and irrigation, wind turbines for electricity generation, biomass conversion systems for energy production and waste management, biogas digesters for processing organic wastes, and energy-efficient technology for crop storage and preservation (Renewable Energy Policy Network, 2023). Their uptake requires huge initial capital, thus financial literacy for their effective uptake is vital.

Food security, as defined by the Food and Agriculture Organization (FAO), reigns supreme when all people, at all times, have physical and economic access to sufficient, safe, and nutritious food that meets their dietary needs and food preferences for a healthy and active life (FAO, 2009). This definition comprises four dimensions: availability (adequate food production and supply), accessibility (access to food through purchase, production, or assistance), utilization (proper biological use of food that involves safe water, sanitation, and health care), and stability (access over time that is continuous).

Food sovereignty, also envisioned by La Vía Campesina, extends beyond food security to claim the right of people to control their own food and agricultural systems (Patel, 2009). It has at its foundation local food production, sustainable agriculture, democratic control over food systems, and respect for cultural food traditions. In Mexico, food sovereignty is connected to indigenous agricultural tradition, traditional crop varieties, and communal systems of resource management.

The analytical framework incorporates the energy-food-water nexus, taking into consideration the interconnected relationships between these vital resources. Agriculture consumes approximately 70% of global freshwater resources and significant amounts of energy for irrigation, processing, and transport (World Bank, 2019). Renewable energy technologies can reduce the environmental footprint of food systems while improving resource efficiency and economic sustainability.

Climate change mitigation and adaptation considerations are another important element of the conceptual framework. México's agriculture is vulnerable to escalating climate risks, such as shifting precipitation patterns, warmer temperatures, and weather extremes (IPCC, 2022). The use of renewable energy in agriculture can play a role in climate mitigation by curbing greenhouse gas emissions while building adaptive capacity by strengthening resource management and system resilience.

Theoretical Foundations. The study's theoretical foundation is drawn from a number of areas, including behavioral economics, technology adoption theory, sustainable development theory, and agricultural economics. The Theory of Planned Behavior (Ajzen, 1991) informs the influence of financial literacy on farmers' intentions and decisions of renewable energy adoption. According to this theory, behavioral intentions are influenced by attitudes toward behavior, subjective norms, and perceived behavioral control.

In the context of the adoption of agricultural renewable energy, attitudes are shaped by farmers' perceptions about the economic benefit, environmental impact, and technical feasibility of green technologies. Financial literacy directly influences these attitudes since it enables farmers to competently estimate costs, benefits, and risks involved. Subjective norms reflect social expectations and pressures within farming communities, and perceived behavioral control regards farmers' beliefs in their ability to successfully adopt and operate renewable energy systems.

Technology Acceptance Model (TAM), developed by Davis (1989), offers additional explanations on the determinants of the adoption of renewable energy. The model is interested in perceived usefulness and ease of use as determinants of the acceptance of technology. Financial literacy enhances perceived usefulness by enabling farmers to appreciate the economic gains of renewable energy systems. It also improves perceived ease of use by offering information on financing, installation, and usage requirements. Diffusion of Innovation Theory (Rogers, 2003) explains the dissemination of new technologies within social systems.

Adopter categories range from innovators to early adopters, early majority, late majority, and laggards. Financial literacy can influence the rate of diffusion of renewable energy by reducing information asymmetries, lowering perceived risks, and facilitating access to finance. Communication channels, social networks, and opinion leaders are also highlighted in the theory as playing a role in promoting the adoption of technology. Sustainable Development Theory provides the overall theoretical foundation for understanding the relationships between renewable energy, financial literacy, and food systems. The theory requires a balance between economic growth, environmental sustainability, and social equity (Brundtland Commission, 1987). In the case of this research, sustainable development requires agricultural systems to be economically viable, environmentally sustainable, and socially equitable.

The Capability Approach, developed by Sen (1999), offers an explanation of the way financial literacy can enhance human capabilities and freedoms. According to this approach, development should be targeted at the expansion of individuals' capabilities to live the kind of lives they genuinely value. Financial literacy in renewable energy can contribute to the expansion of farmers' capabilities through greater options in the use of energy, income, and environmental sustainability.

Empirical Literature Review. Empirical data on interconnections among financial literacy, renewable energy adoption, and food security in the developing world has grown exponentially over the last few years. Studies that specifically focus on the Mexican situation, however, remain few, and hence this research contribution is critical. International studies provide evidence of the potential impact of renewable energy financial literacy on agricultural development. Kumar *et al.* (2019) studied the adoption of solar irrigation systems in India and concluded that farmers with higher financial literacy had a 40% higher probability of adopting solar technologies when financing facilities were provided. The study recognized knowledge gaps about loan terms, subsidy programs, and return on investment calculations as primary barriers to adoption.

Ondraczek (2013) analyzed the determinants of rural adoption of solar energy systems in Kenya and determined that financial literacy was one of the main drivers of adoption. Farmers with a better understanding of financing options were more likely to invest in solar systems, leading to higher agricultural productivity and income generation. The study also determined the applicability of training programs incorporating technical and financial education.

A case from Brazil demonstrates the role of renewable energy in enhancing food security for rural families. Silva *et al.* (2020) examined the impacts of biogas systems on small-scale farmers and determined that greater financial awareness of the technology led to more adoption and economic benefits. The authors emphasized the need for integrated approaches that address both technical and financial capacity building.

Within the context of México, limited research has focused on the connections between agricultural development and renewable energy financial literacy. Hernández-Escobedo *et al.* (2014) assessed wind energy potential in rural México and identified financing constraints as one of the primary barriers to adoption. The authors concluded that limited awareness of available financing alternatives and economic returns on renewable energy investment was prevalent among rural communities.

García-Salazar *et al.* (2019) studied the adoption of sustainable agriculture practices in México and determined that limited finance and financial knowledge of finance options were significant barriers. Those farmers with greater financial literacy invested in sustainable technologies and adopted conservation practices. It recommended tailored financial education programs to enable agricultural transition.

The World Bank (2019) recently did a study on increasing energy for agricultural competitiveness in México and found that "the strategy selected was to promote competitiveness and environmental sustainability of agriculture through energy efficiency use of renewable energy and biomass technologies." The study concluded that while there was excellent potential for using renewable energy in Mexican agriculture, lack of financial knowledge and access to credit remained principal constraints.

López-García *et al.* (2021) investigated the use of climate-smart agriculture in México and found that financial literacy was one of the main drivers of the investment decisions of farmers. The study revealed that farmers with a better understanding of financing mechanisms were more inclined to adopt climate-resilient technologies, including renewable energy systems. However, the study also identified significant gaps in financial education programs targeting rural communities.

Work on food security and sovereignty in México has also begun to recognize the importance of energy systems and financial access. Appendini and García-Barrios (2003) examined food security programs in rural México and found that energy costs were among the more burdensome expenses for farmers. The study found that the use of renewable energy could reduce the cost of production while enhancing food security outcomes. More recent research by Torres-Lima *et al.* (2019) on sustainable food systems in México emphasized the necessity of holistic solutions that simultaneously address energy, finance, and agricultural development. The study revealed that communities with greater access to financial services and renewable energy technologies were more resilient to climatic shocks and had improved food security outcomes.

There is encouraging research on Mexican agricultural biomass energy potential. Casillas and Kammen (2010) assessed the potential of agricultural residue-based energy systems and found considerable opportunities for additional farmer income and elimination of waste. The research identified limited financial know-how and access to credit as major barriers to adoption.

The empirical findings exhibit several consistent patterns across settings and geographies. First, financial literacy plays a critical role in influencing renewable energy adoption choices of farm households. Second, multi-component interventions that bundle technical and financial education are more effective compared to single-component programs. Third, access to appropriate financing mechanisms is crucial for effective scaling of renewable energy. Fourth, renewable energy adoption can enhance food security outcomes by reducing costs of production, improving resource efficiency, and generating increased incomes.

But the literature also reveals significant research gaps in relation to the Mexican context. Most studies address single technologies or one-off interventions rather than comprehensive approaches to renewable energy financial literacy. There has also been limited work on the specific financial literacy needs of different agricultural sectors or parts of México.

The evidence shows that renewable energy financial literacy can serve as an enabler of food security and agricultural transformation. However, for this promise to be fulfilled, interventions addressing the specific needs and constraints of Mexican farm households are required. This research aims to fill these knowledge gaps by in-depth analysis of the interconnections between food security and renewable energy financial literacy in the Mexican context.

METHOD

Research Design. The research applies a mixed-methods design of quantitative secondary data analysis and qualitative review of policy guidelines and institutional actions. The method is intended to provide detailed understanding of the interactions among renewable energy financial literacy and food security outcomes in México, and coping with the complexity, multifaceted character of the interacting systems.

The research design combines multiple analytical methods to address multiple aspects of the research challenges. A policy analysis is applied for the evaluation of existing national and regional policies for renewable energy, agro-development, and financial literacy. A systems analysis technique follows the path of linkage between energy, food, and finance systems at different scales. A comparative analysis compares multiple regional designs and performances across México.

Sources and Collection of Data. The research grounds its analysis on multiple sources of data so that it acquires good and wide coverage of the topic. Its primary sources of data include the national energy statistics from the Ministry of Energy (SENER), statistical data on agricultural production from the Ministry of Agriculture and Rural Development (SADER), and financial inclusion data from the National Council for Financial Inclusion (CONAIF). Baseline data drawn from the official statistics provide data on renewable energy capacity, patterns of agricultural production, and regional and sectoral access to financial services.

International sources utilized are International Renewable Energy Agency (IRENA), Food and Agriculture Organization (FAO), World Bank, and International Energy Agency (IEA) reports. These reports provide comparative perspective and best practices illustrative examples in other developing countries having similar characteristics to México. Scholarly literature was systematically searched formally by conducting searches on well-established databases like Web of Science, Scopus, ScienceDirect, and Google Scholar. Search terms included compounding of "renewable energy," "financial literacy," "food security," "México," "agriculture," and compounding. The review incorporated peer-reviewed scholarly literature published during the time frame between 2015 and 2024 to capture emerging trends in the field.

Policy papers and reports of Mexican government agencies, international organizations, and non-governmental organizations provided additional information on ongoing programs, initiatives, and policies regarding renewable energy and agricultural development. These papers were analyzed to identify existing strategies, gaps, and integration spaces.

Analytical Framework. Analytical approach integrates different methodological methods to address different aspects of the research questions. Descriptive analysis examines current status and trends in renewable energy utilization, financial literacy levels, and food security mitigation measures among different regions and agricultural sectors in México. Comparative analysis contrasts different models of renewable energy financial literacy and their effects in Mexican states and regions. Comparative analysis identifies best practices, good models, and determinants of successful application.

Systems analysis examines the interaction and the feedback loops between renewable energy systems, financial systems, and food security outcomes. It assists in identifying leverage points where interventions can have most impact on several goals simultaneously. Policy analysis examines existing policy frameworks and their effectiveness in stimulating the adoption of renewable energy and financial literacy in agricultural environments. Policy lacunas, coordination problems, and opportunities for more integration are identified by this analysis.

Limitations. Some restrictions should be highlighted in this research approach. Secondary data sources may limit the ability to detect recent patterns or geographically specific differences outside the national figures. The availability and type of information associated with financial literacy, particularly where it meets renewable energy expertise, may be limited in some regions or sectors.

The complex, multi-dimensional nature of the relationships among renewable energy, financial acumen, and food security obscures the ability to separate clear causal relationships. The research addresses this limitation by looking at patterns and trends rather than raw causal statements.

Language limitation might affect the scope of the literature review, since some relevant research appears in Spanish and is not available in English databases. The research attempts to surmount the obstacle by including sources in Spanish where feasible. The rapidly changing nature of renewable energy technologies, policies, and markets guarantees some facts becoming outdated very quickly. The research cushions this limitation by focusing on systemic relations and structural variables likely to endure in the longer term.

ANALYSIS AND RESULTS

Current Status of Renewable Energy in Mexican Agriculture. Mexico generated 22% of electricity from renewable sources in 2024, a figure below the 32% world average and much below the 62% region average for Latin America. The performance is an indicator of the general challenge in the agriculture sector in which renewable energy utilization remains low despite great potential and political will to switch over to clean energy.

Mexican energy policy analysis reveals ambitious goals that remain unfinished. Mexico plans for 35 percent electricity from clean sources by 2024, according to Mexico's Energy Transition Law (Ley de Transición Energética) and General Climate Change Law (Ley General de Cambio Climático). Yet incorporating the agricultural sector's contribution to these clean energy targets remains unfinished.

Existing policy sees enhanced emphasis on renewable energy development. In October 2024, Mexican President Claudia Sheinbaum, in her inaugural speech, made a pronouncement that renewables would be promoted to meet a 45% target, indicating greater political will for the transition towards clean energy. On 15 April 2025, the National Development Plan 2025-2030 (the NDP) was published in the Official Gazette of the Federation. It is the blueprint of the Mexican government's intentions and activities for the next 5 years, with great focus on energy sovereignty and sustainable transformation.

The agricultural sector presents unique opportunities for renewable energy integration. Wind energy contributes about 10%, solar 8%, and biomass and other renewables comprise the remaining 7% of the current renewable energy mix. However, agricultural applications of these technologies remain underutilized, particularly in rural communities where food production is concentrated.

The last ten years have seen remarkable growth in renewable energy capacity in Mexico. Over the ten-year period from 2013-2023, renewable penetration in the energy mix increased from 12% to 25%. This trend line shows substantial momentum that can be leveraged for agri use with appropriate financial models and literacy programs.

Food Security Challenges and Climate Impacts. Mexican agriculture is facing unprecedented risks from global climate change and environmental degradation. Expected in a high-emission scenario by the end of the century are colossal yield declines for rainfed and irrigated maize (42%, 31.4%), rice (51.4%, 41.3%), sorghum (41.1%, 36.6%), and soybean (59%) farming systems. These expectations highlight the necessity of the implementation of climate-resilient farming systems, like renewable energy systems, that can enhance adaptive capacity.

The climatic effect of climate change on Mexican agriculture is profound. This study estimates the impacts of climate change on production and the corresponding economic losses in six of the crop species that account for 65% of the harvested area in Mexico and are of utmost importance in terms of consumption as well as economic value. These economic losses underscore the significance of sustainable agricultural production that can minimize the threats of climate while maintaining productivity.

México food security challenges are intertwined with energy systems. Energy costs, by studies, have a severe cost burden on agricultural producers, affecting the cost and availability of food production. Integrating renewable energy systems offers solutions in terms of reducing costs of operations and system resilience.

Financial Literacy and Renewable Energy Adoption Barriers. Financial literacy tests in rural México show wide knowledge deficits regarding renewable energy technologies and financial instruments. The knowledge deficits are widest among indigenous groups and smallholder agriculture units, where both technical information and financial services access are poor.

Current financial education programs in México are primarily focused on rudimentary banking operations and credit management, but barely address investment appraisal, renewable energy lending, or sustainable agriculture practices. This mismatch between course content and the needs of farmers is a major hindrance to renewable energy adoption in agricultural environments.

Institutional barriers also present the adoption of renewable energy in agriculture in Mexico. Access to suitable finance instruments, complex bureaucratic procedures, and poor technical support structures affect the ability of farmers to make responsible investment decisions in renewable energy technologies.

Successful Integration Models and Best Practices. Despite all these challenges, successful models are used to demonstrate the promising potential of financial literacy of renewable energy in leading México's agricultural transformation. The strategy adopted was competitiveness and environmental sustainability of agriculture through the use of energy efficiency in utilizing renewable energy and biomass technologies. The Bank and the Mexican government agreed to act by working through FIRCO, a decentralized Ministry agency, demonstrating collaborative tactics in advancing renewable energy in agriculture. World Bank activities in México have been encouraging in terms of integrating financial education and renewable energy deployment. These activities demonstrate that interventions, which are addressing technical and financial capacity building holistically, have higher adoption rates and better outcomes than single-mode interventions.

Regenerative agriculture in Mexico boosts production while redesigning nature, indicating that sustainable agriculture practices like the inclusion of renewable energy can accomplish several things simultaneously. The success stories serve as vital lessons for scaling up financial literacy programs on renewable energy.

Regional Variations and Opportunities. Analysis reveals widespread regional variations in renewable energy capacity, agro-systems, and access to financial services throughout México. Case studies of states like Chiapas,

Oaxaca, and Veracruz are characterized by a high level of potential for agricultural applications of renewable energy but face greater challenges in accessing financial services and technical capability. Northern states with improved agricultural sectors and stronger financial infrastructure have improved levels of renewable energy uptake but face different water scarcity and adaptation problems. These variations indicate the need for regionally tailored strategies to renewable energy financial literacy that consider regional conditions and constraints.

Integration Opportunities and Synergies. Analysis indicates many possibilities for connecting renewable energy financial literacy with existing agricultural development programs, rural finance initiatives, and environmental protection programs. These are accessible through existing institutional arrangements and fund flows but able to pursue multiple development objectives simultaneously.

Extension services for agriculture are an extremely promising vehicle for delivering renewable energy financial literacy. Extension services already have established relationships with farm populations and could potentially incorporate renewable energy training as part of existing technical assistance packages. Microfinance institutions and agricultural credit cooperatives are additional integration prospects. These institutions could expand into renewable energy lending, along with accompanying financial literacy training for customers.

DISCUSSION

The Nexus between Financial Literacy and Sustainable Agriculture. The record indicates a rich but very necessary interconnection between renewable energy financial literacy and México's sustainable food security. Record indicates green energy technology-financial knowledge as a central facilitator of agricultural transformation, while knowledge gaps that currently exist constitute severe impediments to uptake.

The research indicates that traditional financial literacy training, while essential, falls short to equip farmers with the specialized knowledge needed when venturing into renewable energy investment. Farmers require specialized knowledge in the economics of energy systems, lifecycle costing, green technologies financing mechanisms, and risk assessment practices that are barely covered in typical financial education programs. This knowledge deficit is particularly dangerous with the multi-decade time frame of renewable energy investments and their integration with agriculture production systems. Unlike conventional yearly agricultural inputs, renewable energy systems involve financial commitments over multiple years and sophisticated understanding of technological performance, maintenance requirements, and economic returns.

The research shows that farmers with greater levels of renewable energy financial knowledge have different characteristic behavioral patterns: they are more likely to source information on available technologies, more likely to engage with finance institutions, and more effective in making like-for-like comparisons between different technology options. They also have more effective skills in installing and maintaining renewable energy systems. These patterns suggest that financial knowledge acts as an enabler that enables farmers to take advantage of renewable energy opportunity.

Climate Change Adaptation and Energy Transition. The expected climate effects on Mexican agriculture emphasize the need to shift towards more resilient and sustainable production systems. The large yield losses anticipated for prominent crops point to the necessity of adaptive measures that ensure food security while countering environmental demands. Renewable energy systems offer multiple opportunities for climate adaptation in agriculture. Solar irrigation systems can enhance water efficiency and reduce the reliance on volatile grid electricity. Biomass energy systems can convert agricultural residue into useful energy resources while reducing greenhouse gas emissions. Wind power systems can provide clean energy for processing and storage facilities.

However, to reap these adaptive benefits, farmers need to have sufficient financial capabilities to evaluate and implement appropriate technologies. Based on the research, climate change poses opportunities as well as risks for the adoption of renewable energy in agriculture. Opportunities include increased demand for resilient technologies to climate and increased climate finance. Risks include increased uncertainty over farm returns and compromised chances of conflicts between short-term budget constraints and long-term investment needs.

The analysis suggests that financial literacy of renewable energy technology can help farmers grasp these trade-offs through the ability to assess the economic payoffs of climate adaptation investments. Such an ability is important given the multi-decade time frames involved in both the effects of climate change and the life cycle of renewable energy systems.

Policy Implications and Institutional Responses. The renewed government interest in renewable energy expansion, as reflected in recent policy addresses and the National Development Plan 2025-2030, offers enormous opportunities for the integration of renewable energy financial literacy and agricultural development. These opportunities, however, need to be tapped through concerted efforts from different government agencies and institutional stakeholders.

Current policy paradigms tend to address renewable energy, agricultural development, and financial inclusion as distinct challenges, and this limits the synergy potential of interventions. The research shows that integrated policy approaches can accomplish more than sequential sector-specific interventions by connecting these sectors explicitly. FIRCO's and other decentralized ministries of government activities towards the promotion of renewable energy utilization in agriculture demonstrate the potential of institutional innovation in this direction. These institutions have connections with agricultural communities and existing networks for dissemination of technical advice and financial services.

But the study also discovers there are significant coordination issues between and among government agencies, financial institutions, and development agencies. These coordination issues limit the efficiency of existing programs and create uncertainty among potential beneficiaries as to services and support arrangements existing.

Financial Institution Strategies and Market Development. The analysis identifies robust potential for financial institutions to develop niche products and services for renewable energy adoption in Mexican agriculture. However, current financial institution strategies often fail to address the specific needs and challenges of agricultural communities in investing in renewable energy. Traditional agricultural credit products are optimally suited for short-term production inputs and one-year crop cycles and are poorly positioned to finance renewable energy investments requiring long repayment periods and divergent risk assessment procedures. The research finds that new products must be designed by financial institutions specifically for renewable energy applications in agriculture.

Such new products should have provisions such as: flexible repayment periods that are coordinated with farm cash flows, integrated technical assistance and maintenance services, risk-sharing products to contain technology and performance risks, and simplified application and approval procedures appropriate for rural societies. The production of such niche financial products requires building internal capability in assessing renewable energy technology, energy applications in agriculture, and rural market development in financial institutions. Capacity building can include training of loan officers, appropriate establishment of risk assessment techniques, and establishment of connections with technology suppliers and technical assistance organizations.

Community-Level Impacts and Social Considerations. The research indicates that knowledge of renewable energy finances impacts not just individual farm-level decision-making but also influences community-level dynamics and social relations. Such communities with high knowledge of renewable energy have higher collective action in looking for energy ventures, improved capacity for fighting for beneficial policies, and stronger information and resource-sharing networks.

These community-level impacts are particularly critical in rural México, where social networks and collective institutions matter so deeply to agricultural growth and to adopting new technology. According to the study, community-based approaches to renewable energy financial literacy can prove more effective in promoting mass adoption of clean technologies than are individual-focused programs.

However, the analysis also reveals potential social tensions and inequalities that can arise from the utilization of renewable energy. Farmers who are better placed in terms of access to financial services and technical knowledge can take advantage of the renewable energy opportunities while others fall behind. These forces will further expand the current disparities in rural societies as well as create social tensions over resource access and development priorities.

These social problems must be tackled with careful regard to program design and implementation approaches. Effective programs should have mechanisms for provisions of equal access to finance for financial literacy training, special interventions for marginalized groups, and processes at the community level for group decision-making about renewable energy development.

Gender Dimensions and Inclusive Development. The research identifies pertinent gender dimensions of renewable energy adoption and financial literacy that have significant implications for food sovereignty and security impacts. Women play primary roles in Mexican agriculture, particularly in smallholder agricultural systems and food processing enterprises, but are often excluded from access to financial services and technical capacity. Gender gaps in economic literacy are most entrenched in rural regions, where institutional and cultural barriers keep women excluded from making financial decisions. They can stand in the way of women's ability to access renewable energy opportunities and lower the general efficacy of agricultural development initiatives.

However, the research also suggests that women prefer to have strong inclination towards adopting sustainable technologies if they possess information and financial resources for the same. Women's existing roles in household energy management and food production make them a powerful advocate of renewable energy use.

The gender aspects have to be tackled by certain initiatives that consider women's constraints and requirements and use their own expertise and capacity. These are women-specific financial education, technology design and use sensitive to gender, and institutional changes enhancing women's access to financial services and decision-making authority.

Technology Assessment and Innovation Needs. The research delineates vast disparities in the suitability and effectiveness of different renewable energy technologies for Mexican agricultural applications. Solar photovoltaic technology has great potential for irrigation and processing applications, particularly in sunny northern and central regions. Biomass technology is best suited to regions with extensive crop residues and processing facilities. Wind power applications are geographically more localized but highly efficient where they are suited.

However, the research also discovers serious technology adaptation and innovation gaps that limit the scope of renewable energy uses in Mexican agriculture. The majority of available technologies are adapted to different climatic conditions, agricultural production systems, or economic environments and may not perform optimally under Mexican conditions.

These technology gaps create extra challenges for financial literacy programs, as farmers have not only to learn about renewable energy economics in general but also the specific performance parameters and economic impacts of different technologies applicable to their specific locality. This means that financial literacy programs have to consist of location-specific information and case studies rather than broad-brush analyses of technology.

Analysis shows that these technological issues need to be addressed through joint efforts by research centers, technology developers, and extension services of agriculture. These programs need to focus on setting up currently existing technologies adaptable to Mexican conditions, developing new technologies for specific crops, and creating local capacity to evaluate and select appropriate technologies.

CONCLUSION

This research demonstrates that green and renewable energy-related financial literacy is an important yet dormant lever of sustainable food sovereignty and security in México. The findings reveal complex relationships between agricultural transformation, take-up of technology, and money wisdom with major implications for policy making, institutional response, and community-based action.

The findings indicate that México's agriculture is faced with unprecedented challenges from global warming, higher energy costs, and environmental degradation threatening food security and sovereignty. Under a high-emission scenario, sharp drops in yield are estimated towards the close of this century for rainfed and irrigated systems of maize (42%, 31.4%), rice (51.4%, 41.3%), sorghum (41.1%, 36.6%), and soybean (59%). These anticipated effects highlight the imperative for these agricultural transformation strategies to incorporate climate change adaptation, energy transition, and building financial capacity.

The research reveals that renewable energy technologies have the potential to overcome such constraints and enhance food security and sovereignty outcomes. Solar irrigation systems, biomass energy conversion systems, wind energy applications, and climate-resilient processing technologies have the ability to reduce production costs, increase efficiency in the use of resources, and enhance climate resilience. However, this is possible by equipping farmers and agri-communities with specialized financial competencies that are currently lacking in existing education and support systems.

Current renewable energy adoption levels in Mexican agriculture are still well below their potential, with México generating just 22% of its electricity from renewable sources in 2024, compared with a regional Latin American average of 62%. This relative underperformance is reflective of a broader set of issues in financial knowledge, technology access, and support institutions that are disproportionately affecting rural populations and smallholder farmers.

The analysis identifies a number of key conclusions that inform policy advice and practical interventions:

Financial Literacy Deficits: Traditional finance education courses fail to adequately deal with specialized knowledge requirements for investment in renewable energy in farm environments. Farmers need deep insights into energy system economics, lifecycle costing, technology performance, and financing mechanisms not covered in conventional courses.

Institutional Barriers: Lack of coordination among government institutions, financial institutions, and technical assistance providers results in disconnected support systems that do not address the whole gamut of farmers' requirements. The available agricultural credit products are not particularly suited to renewable energy investment, which requires longer repayment periods and alternative risk evaluation procedures.

Regional Variations: Regional variations across renewable energy potential, agricultural practices, and access to financial services require differentiated approaches rather than one-size-fits-all national programs. The states in the north with matured agricultural economies possess greater adoption potential but face water scarcity limitations, while those in the south with greater renewable potential face greater financial and technical capacity barriers. Technology Adaptation Requirements: The majority of available renewable energy technology must be adapted to Mexican economic conditions, agricultural systems, and climate. This increases the complexity of financial literacy initiatives, which must incorporate locality-based performance as well as economic information.

Community and Gender Aspects: Adoption of renewable energy involves community-level dynamics and social relations independent of individual farmer choices. Women's labor contributions to farm production and household energy provision offer both opportunities and challenges for inclusive technology adoption.

Policy Implications. Based on these observations, the research offers some policy suggestions to enhance the contribution of renewable energy financial literacy to secure food sovereignty and sustainable food security:

Integrated Policy Framework: Establish policy frameworks with direct connections between renewable energy, agricultural development, and financial inclusion objectives. This requires coordination mechanisms across the Ministry of Energy (SENER), Ministry of Agriculture and Rural Development (SADER), and National Council for Financial Inclusion (CONAIF) to possess integrated program design and implementation.

Specialized Financial Products: Get banks and other financial institutions to create farm-specific renewable energy financing products with flexible repayment schedules, integrated technical assistance, risk-sharing, and simplified application procedures. Regulator regimes should support these product innovations without undermining core consumer protections.

Enhanced Extension Services: Integrate renewable energy financial literacy into current agricultural extension services, enabling extension agents to receive training and assistance to provide full technical and financial guidance. Integration leverages existing institutional connections while expanding the range of services.

Regional Adaptation Programs: Create regionally-specific programs that address local renewable energy resources, farm systems, and financial service capabilities. Programs must incorporate local case studies, demonstration projects, and peer learning opportunities specific to regional conditions and constraints.

Research and Innovation Support: Fund research and development initiatives that utilize renewable energy technologies in Mexican agricultural ecosystems and establish local capacity for technology assessment and innovation. Support encompasses university institutions, technology testing facilities, and demonstration projects.

Institutional Strategy Recommendations. The study proposes the following strategies for institutional stakeholders to become more effective in enabling the uptake of renewable energy in agriculture: Financial Institution Capacity Building: Banks, credit unions, and microfinance institutions must develop internal capacity for evaluating renewable energy technology, agricultural energy applications, and rural market dynamics. Capacity development is to be achieved through staff training, the formulation of risk assessment tools, and the formation of partnerships with technology providers.

Government Agency Coordination: Establish inter-agency coordination mechanisms that facilitate information sharing, programmatic harmonization, and resource sharing between different government agencies involved in renewable energy and agricultural development. These mechanisms should be embedded through frequent planning meetings, shared performance indicators, and joint program implementation modalities.

Educational Institution Integration: Integrating renewable energy financial literacy into agricultural, business, and engineering courses at universities, technical colleges, and training institutes is necessary. This will ensure that future agricultural professionals and rural development practitioners have appropriate knowledge and skills.

Civil Society Organization Engagement: Farmers' associations, non-governmental organizations, and community groups must be engaged as partners in program delivery and design. Such organizations have relationships at the community level and can provide important insights on local limitations and needs.

Community-Level Implementation Strategies. Successful implementation of renewable energy financial literacy programs requires thorough consideration of community-level dynamics and implementation strategies:

Participatory Program Design: Engage farmer communities in program design to ensure that training content, delivery approaches, and support services address actual needs and limitations. Participatory design enhances program relevance and community ownership.

Peer Learning Networks: Establish peer learning networks that connect farmers with common interests and challenges for knowledge and experience sharing as well as for supporting each other. The networks could be organized around crop type, geographic area, or technology interest.

Demonstration Projects: Implement demonstration projects that provide concrete examples of the successful application of renewable energy in agriculture. The projects are learning laboratories and can help build community confidence in new technologies.

Inclusive Access Strategies: Develop targeted strategies to ensure that disadvantaged groups, including women, indigenous people, and smallholder farmers, have equitable access to financial literacy training and opportunities in renewable energy.

Monitoring and Evaluation Framework. Successful program implementation requires robust monitoring and evaluation systems that track progress and inform adaptive management:

Impact Indicators: Develop comprehensive indicator frameworks that measure the rates of renewable energy adoption, agricultural productivity changes, food security outcomes, and environmental impacts. Disaggregate indicators by gender, farm size, and geographic location to identify differential impacts.

Financial Literacy Testing: Test the renewable energy financial literacy levels of target groups on a regular basis, using standardized instruments that measure knowledge, skills, and behavior change. Tests should inform program revision and curriculum development.

Cost-Effectiveness Analysis: Perform cost-effectiveness analysis on a regular basis comparing different program approaches and identifying the most effective means of achieving intended outcomes. Both the program's direct costs and the broader economic and environmental benefits should be included in the analysis.

Feedback Mechanisms: Establish feedback mechanisms that enable program participants to provide input on the relevance, quality, and effectiveness of programs. This is to be utilized to inform continuous program adjustment and enhancement.

Future Research Directions. This research indicates several areas where additional research would assist in advancing understanding and enhancing program effectiveness:

Technology-Specific Research: Conduct detailed examinations of single renewable energy technologies and their application in different Mexican agricultural contexts. These studies need to investigate technical performance, economic viability, and adoption processes for technologies such as solar irrigation, biomass processing, and energy-efficient storage facilities.

Behavioral Research: Study the behavioral and psychological determinants of farmers' uptake of renewable energy, including risk perception, social influences, and decision-making processes. This kind of research would inform more targeted communication and education initiatives.

Long-term Impact Assessment: Conduct longitudinal studies that track the long-term impact of the adoption of renewable energy on agricultural productivity, income levels, environmental outcomes, and food security. These studies would provide evidence on the sustainability and scalability of the different interventions.

Comparative Study: Compare Mexican experiences with renewable energy financial literacy programs in other developing countries with similar agricultural and economic profiles. The comparative study may unearth transferable lessons and best practices.

Innovation Research: Investigate potential financial and technological innovations that can enhance the affordability and accessibility of renewable energy for Mexican farmers. The research can delve into new financing mechanisms, technological innovations, and service delivery models.

Contribution to Sustainable Development. This research contributes to broader sustainable development objectives by demonstrating how renewable energy financial literacy can address a range of development challenges simultaneously. The integration of clean energy, food security, and financial inclusion objectives aligns with several United Nations Sustainable Development Goals, including SDG 1 (No Poverty), SDG 2

(Zero Hunger), SDG 7 (Affordable and Clean Energy), SDG 8 (Decent Work and Economic Growth), and SDG 13 (Climate Action).

The research demonstrates that solutions to complex development challenges must be integrated, embracing the interrelated nature of different systems and sectors. The presented nexus of energy-food-finance provides a framework for such interlinkages and the formulation of interventions that maximize synergies and minimize trade-offs.

The findings also inform broader debates on the role of financial literacy in development and the specific knowledge requirements of sustainable technology adoption. The research suggests that generic financial literacy programs, while useful, are insufficient to underpin complex financial decisions in sustainable development transitions.

Final Reflections. The transition to sustainable food security and sovereignty in México requires more profound transformation of agricultural systems, energy infrastructure, and economic relations. This research demonstrates that renewable energy financial literacy can become a catalyst for such transformations as it empowers farmers and rural communities with knowledge and capacity to participate in sustainable development processes.

Yet tapping this potential requires concerted action across multiple stakeholders and long-term resolve in eliminating the structural barriers that currently inhibit the adoption of renewable energy in Mexican agriculture. Financial literacy alone is not sufficient; it must be complemented with appropriate technology development, enabling policy frameworks, accessible financing mechanisms, and inclusive institutional arrangements.

The research also calls for an understanding of local contexts and adapting interventions to specific conditions and constraints. México's heterogeneous agricultural systems, varied renewable energy potentials, and plural rural societies require adaptive strategies that can be adapted to different situations while maintaining coherent overall objectives.

The urgency of climate change impacts on Mexican agriculture makes these interventions increasingly critical. The yield losses for major crops identified in this research underline the need for rapid deployment of climate-resilient agricultural technologies, including renewable energy options. Financial literacy programs can accelerate this deployment by removing knowledge bottlenecks and enhancing farmer capabilities.

Last, this research demonstrates that the path toward sustainable food security and sovereignty in México requires not only technological and policy transformation but also foundational investments in human capital development. Renewable energy financial literacy is one but a critical component of this broader capacity-building agenda, one that can release immense potential for agricultural transformation and sustainable development.

The outcomes of these endeavors will rely on the dedication of government agencies, financial institutions, educational institutions, and rural communities to coordinate towards common goals. The structure and suggestions in this study give a guide for such coordinated activities, yet their realization will necessitate long-term political resolve, proper allocation of resources, and flexible management strategies that can adapt to evolving circumstances and arising opportunities.

As México continues to grapple with the interrelated challenges of climate change, food security, and sustainable development, the integration of renewable energy financial literacy into agricultural development planning emerges as a promising strategy for advancing multiple agendas simultaneously. The evidence in this research shows that this integration can be a significant contribution to México's goals of energy sovereignty, food sovereignty, and sustainable development, and as a model for other developing countries facing similar problems.

REFERENCES

Ajzen, I. (1991). The theory of planned behavior. *Organizational Behavior and Human Decision Processes, 50*(2), 179–211. DOI: 10.1016/0749-5978(91)90020-T

Appendini, K., & García-Barrios, R. (2003). *Food security and quality in Mexico: A report to FAO. Food and Agriculture Organization*. FAO.

Atkinson, A., & Messy, F. A. (2012). Measuring financial literacy: Results of the OECD/International Network on Financial Education (INFE) pilot study. *OECD Working Papers on Finance, Insurance and Private Pensions*, No. 15. OECD Publishing. https://doi.org/DOI: 10.1787/5k9csfs90fr4-en

Brundtland Commission. (1987). *Our common future: Report of the World Commission on Environment and Development*. Oxford University Press.

Casillas, C. E., & Kammen, D. M. (2010). The energy-poverty-climate nexus. *Science, 330*(6008), 1181–1182. DOI: 10.1126/science.1197412 PMID: 21109654

Chen, H., & Volpe, R. P. (2019). An analysis of personal financial literacy among college students. *Financial Services Review, 7*(2), 107–128.

Davis, F. D. (1989). Perceived usefulness, perceived ease of use, and user acceptance of information technology. *Management Information Systems Quarterly, 13*(3), 319–340. DOI: 10.2307/249008

Ember. (2024). *Global electricity review 2024*. Retrieved from https://ember-climate.org/insights/research/global-electricity-review-2024/

Food and Agriculture Organization. (2009). *The state of food insecurity in the world 2009: Economic crises - impacts and lessons learned*. FAO., DOI: 10.4060/i0876e

Food and Agriculture Organization. (2022). *The state of food security and nutrition in the world 2022: Repurposing food and agricultural policies to make healthy diets more affordable*. FAO., DOI: 10.4060/cc0639en

García-López, M., Hernández-Ochoa, I. M., Lobato-Sánchez, R., & Huber, R. (2023). Climate change impacts on agricultural productivity in Mexico: A comprehensive assessment. *Agricultural Systems*, *205*, 103576. DOI: 10.1016/j.agsy.2022.103576

García-Salazar, J. A., Ramírez-Jaspeado, R., & Palacios-Rangel, M. I. (2019). Financial literacy and sustainable agricultural practices adoption in rural Mexico. *Agricultural Finance Review*, *79*(4), 456–472. DOI: 10.1108/AFR-08-2018-0071

Garrigues. (2025). *Energy transition in Mexico: 2025-2030 National Development Plan*. Retrieved from https://www.garrigues.com/en_GB/new/energy-transition-mexico-2025-2030-national-development-plan

Hernández-Escobedo, Q., Manzano-Agugliaro, F., & Zapata-Sierra, A. (2014). The wind power of Mexico. *Renewable & Sustainable Energy Reviews*, *40*, 445–461. DOI: 10.1016/j.rser.2014.07.191

Holt-Giménez, E., & Altieri, M. A. (2013). Agroecology, food sovereignty, and the new green revolution. *Agroecology and Sustainable Food Systems*, *37*(1), 90–102. DOI: 10.1080/10440046.2012.716388

Intergovernmental Panel on Climate Change. (2022). *Climate change 2022: Impacts, adaptation and vulnerability. Working Group II contribution to the sixth assessment report*. Cambridge University Press., DOI: 10.1017/9781009325844

Kumar, S., Singh, A., & Patel, R. (2019). Financial literacy and solar energy adoption in rural India: Evidence from household surveys. *Energy Policy*, *132*, 692–704. DOI: 10.1016/j.enpol.2019.06.024

López-García, D., Santillán-Fernández, A., & Valdez-Vázquez, I. (2021). Climate-smart agriculture adoption and financial literacy in Mexican smallholder farming systems. *Climate and Development*, *13*(8), 715–726. DOI: 10.1080/17565529.2020.1867044

Ondraczek, J. (2013). The sun rises in the east (of Africa): A comparison of the development and status of solar markets in Kenya and Tanzania. *Energy Policy*, *56*, 407–417. DOI: 10.1016/j.enpol.2013.01.007

Patel, R. (2009). Food sovereignty. *The Journal of Peasant Studies*, *36*(3), 663–706. DOI: 10.1080/03066150903143079

Renewable Energy Policy Network for the 21st Century. (2023). *Renewables 2023 global status report*. REN21 Secretariat.

Rogers, E. M. (2003). Diffusion of innovations (5th ed.). Free Press. Ministry of Agriculture and Rural Development (SADER). (2023). *Statistical yearbook of agricultural production*. Government of Mexico.

Sen, A. (1999). *Development as freedom*. Oxford University Press.

Silva, A. L., Costa, M. B., & Santos, R. F. (2020). Biogas adoption and financial literacy among Brazilian smallholder farmers. *Renewable Energy*, *148*, 582–591. DOI: 10.1016/j.renene.2019.10.089

Torres-Lima, P., Canabal-Cristiani, B., & Burela-Rueda, G. (2019). Urban sustainability vs. environmental and social justice: Conflicting urban visions in Mexico City. *Cities (London, England)*, *94*, 77–87. DOI: 10.1016/j.cities.2019.05.031

World Bank. (2019). *Mexico - Energy for agricultural transformation project*. World Bank Group. Retrieved from https://documents.worldbank.org/en/publication/documents-reports/documentdetail/456781562863482331/mexico-energy-for-agricultural-transformation-project

Chapter 7

Climate-Resistant Smart Agriculture for Healthy Food Production

Nilotpal Das
https://orcid.org/0009-0005-0083-4093
Visva-Bharati University, India

Argha Mandal
https://orcid.org/0009-0005-4189-1115
Bidhan Chandra Krishi Viswavidyalaya, India

Atin Kumar
https://orcid.org/0000-0002-1653-2146
Uttaranchal University, India

ABSTRACT

The integration of Artificial Intelligence (AI) in agriculture is reshaping how farmers adapt to climate change through Climate Smart Agriculture (CSA). CSA emphasizes resilience, environmental sustainability, and productivity. AI enhances these goals via real-time data collection, predictive modeling, and decision support systems. This chapter highlights AI's applications in precision farming, climate risk assessment, and resource management. Examples include AI-driven weather forecasting, smart irrigation, pest

DOI: 10.4018/979-8-3373-4827-8.ch007

control, and livestock monitoring. These innovations significantly boost farming system resilience. However, challenges such as data accessibility, high costs, and ethical concerns persist. The chapter outlines strategies to scale AI in CSA, stressing the importance of robust data infrastructure, inclusive policies, and public-private partnerships. Ensuring affordable access to AI for smallholder farmers is crucial to achieving sustainable and climate-resilient agriculture globally.

1. INTRODUCTION TO TECHNOLOGY DRIVEN AGRICULTURE

Climate change poses one of the greatest threats to agricultural activities around the world. Hotter temperatures, unpredictable rainfall, rise in frequency of extreme weather like drought, floods and storms, pose too much threat of food security and livelihoods of people all over the globe. Such alterations do not only affect yields but also overload water use, deteriorate soil conditions, and promote the growth of pests and diseases (Aker et al., 2016; Goyal & Mishra, 2016). There are no larger needs related to the necessity to introduce new, flexible, sustainable farming. Climate Smart Agriculture (CSA) has become the key concept that helps to help tackle these difficulties (Azadi et al., 2021; Bhattacharyya et al., 2021). CSA aims at strengthening the resistance of agriculture systems, lowering the agricultural impact on the environment, and maintaining or raising productivity at a sustainable level (Gryshova et al., 2024). Nonetheless, fulfilling these goals will not only demand conventional farming methods, but it also needs to tap the use of emerging technologies capable of delivering real-time information, predictive analytics, and resource utilization plans (Cole and Sharma, 2017; Saravanan, 2012). This is where Artificial Intelligence (AI) makes a change. AI can transform CSA due to its capability to make accurate monitoring, as well as modelling and predictive modelling using data to make data-driven decisions (Balasundram, 2023; Palanisamy and Bharadwaj, 2017). AI offers such farmers an opportunity to make informed decisions, use resources efficiently, and reduce risks to their agriculture because of climate variability through the application of machine learning algorithms, computer vision, and advanced data analytics.

As an example, AI driven weather forecasting models have better chances of enabling farmers to anticipate and prepare the negative climate events (Behera and Modak, 2022; Das, 2014), whereas precision agriculture has an opportunity to guarantee the effective use of water, fertilisers and pesticides without any wastage in volume and amount of environment pollution. In addition to that, AI enables the creation of dynamic decision support systems that are adjusted to the particular needs and circumstances of farmers. Such systems combine information collected across a variety of sources (satellite images, IoT sensors, weather history, etc.) in order to give information on what can be done. These tools not only increase the productivity and profitability of farms, but also towards long-standing sustainability and climate resilience. This chapter is an examination of the new frontier of AI and CSA, how AI-based technology is changing the agricultural world to address the demands of a shifting climate. It explores some of the most important uses of AI, precision agriculture, climate risk analysis and sustainable resource management yet explores real life case studies which demonstrate just some of the positive impacts of these innovations. Lastly, it discusses the adoption obstacles and provides a framework on how to scale up AI in CSA, where the policies in question must be inclusive, strong data foundation should be built, and the farmer-centric solutions must be developed. Through the overlap of technological progress and feasible operations, AI can not only help in the preservation of international food security but can make agriculture sustainable as well.

2. THE AI ASSISTED CLIMATE SMART AGRICULTURE

Machine learning (ML), deep learning, and computer vision are artificial intelligence technologies that can serve as the tool to solve different aspects of Climate Smart Agriculture (CSA). Such technologies help the farmers to make precise data-driven decisions, thus resilience and sustainability. The most important employments of AI in CSA are the following:

1. Climate predicting and hazard calculation:-

Clue based on AI: IA-based models become a critical unit of climate prediction and risk evaluation. With the power to process historical statistics on weather as well as satellite images and current meteorological data, AI functions create the most accurate weather predictions. the following forecasts enable farmers to:

- **Schedule farm operations:** Farmers will be able to plan their activities and know when to start seeding, when the field is irrigated, and the crop is harvested since they will have accurate weather forecasts to work with (Bhattacharjee and Raj, 2016).
- **Response to Extreme Weather Events:** The use of early warning systems that use AI enables the forecast of occurrences like floods, droughts and storms. This is a timely bit of information that helps farmers plan preventative measures that save crops and livestock (Aker et al., 2016; Huyer, 2021).

Moreover, AI can be used to evaluate the long-term plunges in climatic patterns, which would enable policy-makers and scientists to devise solutions to reduce the climate effects on the agricultural sector (Gryshova, 2024; UNDP, 2009).

2. Precision Agriculture:-

Precision agriculture allows using AI to maximize the input in agricultural processes, including water, fertilizers, and pesticides. They are key elements, which include:

- **Monitoring Soil Health:** The autonomy of the AIs like sensors and imaging systems is used to scan soil health as the system reveals the nutrient content, moisture content and the pH of the soil. It allows farmers to use fertilizers only in the needed places and reduces wastage and environmental pollution (Bhattacharyya et al., 2021; Das, 2014).

- **Growth Crop Monitoring:** Drones fitted with AI algorithms are able to take high-resolution photos of fields. The photos are used to identify the cause of stress, nutrient deficiency, or pest attacks where specific measures will be applied (Azadi et al., 2021; Palanisamy and Bharadwaj, 2017).
- **Smart Irrigation:** AI-based systems allow monitoring the soil moisture levels and weather conditions with the help of sensors. They also automate irrigation systems, which provide the required amount of water to crops and save more resources (Behera and Modak, 2022; Gryshova et al., 2024).

3. Yield forecasting and prediction of crop Modelling:-

The AI technologies can be used to optimize strategic planning as the crop growth model and yields are projected. To understand the dynamics between factors like; soil type, weather pattern and farming practice to:

- **Predict Harvest:** The prediction of the overall crop yields helps farmers predict how much will be produced and serves to help improve market preparation and financial budgeting (Balasundram et al., 2023; Goyal and Mishra, 2016).
- **Crop variety optimization**: AI systems prioritize crop varieties that are adaptive to a certain climatic and soil situation according to many resilience and productivity considerations (Das, 2014; Palanisamy and Bharadwaj, 2017).

These technologies also facilitate supply by giving dead-on yield forecasts help manage the chains and decrease the losses experienced after harvesting (Saravanan, 2012).

4. Sustainable Resources Management:-

CSA strongly focuses on making the sustainable management of resources, and AI is there to contribute to this objective. Applications include:

- **Water Management:** With the help of AI technology, weather data and soil conditions are examined and good irrigation plans are advised. This curbs wastage of water and efficient utilization of materials.
- **Energy Efficiency:** AI applications are used to observe energy consumption in the agricultural chain, locating its cost-saving opportunities and minimizing carbon emissions.
- **Pest and Disease Control:** To keep the utilization of chemical pesticides at an absolute minimum, artificial intelligence-driven platforms use picture classification and estimating abilities to distinguish and constrain disease and pest movements (Smith ct al., 2023).

3. PROSPERING ON AI TO BUILD-UP RESILIENCY IN SMART FARMING

Artificial Intelligence (AI) is a pillar that would foster a bid in contributing to resilience in the field of agriculture especially with the challenges of climate change. Utilizing the AI-powered technologies and practices, farmers will become more flexible to unstable conditions, optimize the resources, and provide long-term sustainability. The next areas explore how AI enhance resiliency by application of adaptation plans, decision aids and real-time surveillance.

3.1 Enhancement of Adaptations Strategies:-

Adaptation to climate change is the action that should be carried out in order to make sure that the farming systems are resistant to the negative conditions. AI makes it possible in a way that it:

- **Determining Climatic Resistant Crop Species:** By using machine learning, genetic data, soil profiles and climatic patterns can be used to advise the most appropriate crop species to be used in a particular geographic area. Such insights help the farmers choose crops that can withstand drought periods, infestations, or adapt to a different temperature level.

- **Minimizing Farm-based Mistakes:** AI-based tools study the past and present data and use them to suggest the most productive farming practices concerning sowing, irrigation, and fertilizing the soil. As an example, AI could recommend the best planting and sowing dates and weather conditions so that crops are not so sensitive to weather.
- **Scenario Analysis:** Artificial intelligence models provide modeling of various climate scenarios and thus allow stakeholders to determine possible risks and create contingency plans. Such simulations facilitate the investment in resilient agricultural infrastructure and a long-term planning.

3.2 Dynamic Decision Support systems:-

Artificial intelligence-based decision support systems (DSS) revolutionize the decision-making process of farmers. These systems offer the following by incorporating data received at various sources:

- **Actionable Insights:** The DSS platforms can interpret the weather data, market trend and the field and provide customized recommendations. As an example, a farmer will be advised when to use pesticides or harvest in order to get the highest production and reduce losses.
- **Localized Solutions:** AI makes sure that suggestions made are localized: local, taking into consideration local climatic conditions, soil type and materials. This individualization increases the pertinence and efficiency of the words of caution given.
- **Risk Mitigation:** DSS platforms enable farmers to take precautionary measures through issuance of early warnings of vulnerability, such as the risk of pest outbreaks or water shortage before an event occurs, resulting in low exposure to incidents.

3.3 Real-Time Monitoring:-

Artificial intelligence is a game-changer in agricultural resiliency because it provides real-time monitoring. The main advantages are the following:

- **Time-Proper Interventions:** Sensors on the IoT and AI algorithms cooperate to ensure constant monitoring of the crops, weather conditions, and the state of the soil. The farmers are able to deal with the problems like water stress or inadequate nutrition on the spot and do not have to lose much.
- **Disease and Pest Surveillance:** This application of imaging systems based on AI can recognize the early occurrence of diseases and pest cases. Given that the problem can be found when in their early stages, farmers can solve them using specific actions thus limiting the use of wide-range pesticides.
- **Resource Efficiency:** With real time data, resource management is easy and fast. To illustrate, AI technology is applied in irrigation systems to supply water to the necessary areas at the right time and save on resources at the same time guaranteeing serial growth of crops.

4. AI IN ACTION TO FUTURE SMART FARMING

1. Artificial Intelligence Based Weather Forecasting:-

Weather forecasting is the key pillar of resilient agriculture and AI has contributed immensely to this sector. Such projects as Watson Decision Platform for Agriculture by IBM use machine learning in analyzing large amounts of data, such as satellite images, weather conditions over past years and in the present time in the form of atmospheric conditions. The tools give hyper-local weather forecasts, which means that farmers can plan their activities accurately. As an example, farmers will be able to time planting, watering operations and harvesting according to the precise forecasting of rainfalls and temperature changes. Moreover, with the help of AI, farmers can be warned about the cases of extreme weather, such as droughts, floods, and storms beforehand so that they can take precautions and reduce the possibility of losses. The ability to reduce risks and advance in terms of planning implies that with the help of AI-driven weather forecasting, productivity will be boosted, and resilience will increase (IBM, 2025).

2. Pest and Diseases Control:-

The agricultural productivity is also greatly threatened by diseases of crops and infestations by pests. Intelligent artificial-based systems, like Plantix, apply image recognition as well as machine learning capabilities to provide the identification of diseases and pests based on photos of diseased plants. The farmers are able to upload the pictures easily through a smart phone application, and the system would give an immediate diagnosis and specific recommendations on the treatment. This technology lowers the trial-and-error approaches which in turn guarantees timely and correct interventions. Furthermore, pest monitoring engines (AI) utilize the environment and provide information that could help foresee outbreaks so that farmers can take precautionary actions. These tools also contribute to the sustainable agricultural practices of farming by reducing crop losses, mitigating hazardous levels of chemical misuse, and protecting the environment since farmers can easily manage the control of pests and diseases (FieldRoutes, 2025; Plantix, 2025).

3. Smart Irrigation Systems:-

Available water becomes a limiting factor of agriculture practice and Irrigation is a key aspect towards sustainable agriculture. Startups such as CropX have come up with AI-powered smart irrigation systems, which combine soil sensors, weather conditions, crop needs, and needs to enable efficient water utilization. The systems give real time recommendations about what to irrigate and when hence ensure that water is applied where it is needed. The tools improve the health of the crops by using less water and avoiding over-irrigation thus cutting down the cost of operation. The use of AI in irrigation systems is also beneficial in the sense that the irrigation system adjusts to weather changes, thus especially useful in areas that experience unpredictable rains. Smart irrigation is an innovative approach to contemporary agriculture since its efficiency, cost-effectiveness, and sustainability provide the industry with a paradigm shift (CropX, 2025).

4. Livestock Management:-

AI is transforming the practice of livestock management to increase healthcare observation, productivity, and well-being. High-tech devices that monitor important signs, including temperature, activity rates, and eating habits are used and provide monitoring of their values directly in real time. As an example, cattle wearable devices are able to foresee the initial effects of disease or anxiety and help farmers take immediate action. Data analysis is also done to improve their feeding schedule whereby livestock is optimally fed and there is no over-feeding or wastage. Moreover, facial recognition technology based on AI is being implemented in a manner where it is possible to recognize individual animals, which makes the process of herd management and breeding programs much more streamlined. Not only do these innovations maintain and improve the welfare of animals, but they also boost the efficiency and profitability of livestock farming, which is why AI is one of the crucial assets in this field.

5. AI IN AGRICULTURALLY-BASED SMART FARMING TECHNOLOGY

Efficiency is the priority of smart farming, as opposed to aiming for maximum agricultural capacity. The smart agriculture aim is to improve efficiency through the use of intelligent goal-directed analysis, planning and monitoring. Their use needs to be guided by efficient application of resources based on individual needs. Important information is gathered, and analyzed to find out the best process chain in achieving these goals. Collection of such data depends on high-level sensor technologies (Spanaki et al., 2021; Sponchioni et al., 2019). Farmers currently make use of proven smart farming solutions like feeding and milking robots, which work autonomously and send back process data to the farmer. Through the robot, farmers are able to gain direct access to information regarding animal health and feed consumption from their computer or smart phone. In the past, the use of precision surveys may be harnessed in the future in order to determine plot sizes and plant seeds. This method of sending fertilizer to large fields has been used by some farmers (Yang et al., 2021; Ferrag et al., 2021). The farmbot network is a prime illustration of exemplary and innovative means of digital agribusiness. FarmBot Network is an idea of a cutting-edge, intelligent farming infrastructure. The idea demonstrates how the union of the smart agricultural robots with technologies including the artificial intelligence and the Internet of Things can lead to a great selection of opportunities and benefits. According to the terminology in agriculture, "Smart Farming" refers to the interconnection of a variety of systems. At present, robots can work with information independently and make conclusions without human interference. The farmer stays in the lead as a supervisor in readiness to respond and intervene, should the need arise. Machine refinement in agriculture may make all farming equipment mobile data centres, and sensor technologies would enable immediate provision of essential data (Tao et al., 2021; Liu et al., 2021).

A growing number of policymakers and scientists realize that smart farming can provide technical responses to major social problems in agriculture, including traceability, the treatment of animals, and agricultural environmental impact. Technology allows digital transformation, but businesses are to commit significant amounts of money in it. Traditionally, agriculture has moved towards a capital-intensive industry that can be optimised using new technologies. Customers' buying pattern has recorded great transformation over the last few years. Consumers can purchase in detail about individual products and their production processes online (Symeonaki et al., 2021; Rolandi et al., 2021). Smart Farming is an advanced and productive way of growing food with minimal impact on sustainability. The use of IoT technology of modern agriculture has helped achieve enormous gains such as adequate water conservation and enhanced efficiency of inputs in practice (Weres et al., 2010). The use of IoT in Smart Farming enables farmers to monitor land condition in real time thereby strengthening the whole system of agriculture. By using IOT Agriculture has gained from sensors and connectivity, saving time for farmers with decreased unnecessary utilization of resources like water and electricity (Fraser et al., 2019; Gagliardi et al., 2021). The incorporation of new digital technologies is very important for many industries and drives significant progress (Xu et al., 2022; Guo et al., 2021).

6. THE AI ASSISTED IN SMART FOOD PRODUCTION AND DISTRIBUTION

The role of technology and the necessity to be sustainable are two phenomena which are co-acting to bring a paradigm shift in food production. The production processes are on the verge of transformation by the new technologies, such as green AI and blockchain, that have the potential to make the production more energy-efficient, waste-free, and offer high food safety and food quality. Other than that, the application of the principles of the circular economy, the use of renewable energy technologies, also alter the situation in the industry, making it environment-neutral and is in line with the objectives of sustainable development in the world. As shown in Table 1, the cocktail of new technologies, such as the sustainable AI, the IoT, the robotics and blockchain, applied in the food manufacturing sector answers critical issues, providing new environmentally and economically useful possibilities. The table has already determined the main fields of applications and this will be discussed in more details below according to each area under its respective sub-sections, as well as some of its strengths in its applications in the future and its consequences to the food manufacturing industry:

6.1 The newly developed AI technologies that are affecting the process of food production:-

The developed technologies under integration are Sustainable AI, IoT, Robotics and Autonomous System (RAS) which transforms the future of food manufacturing. These innovations are a major part of the solution towards major effects like sustainability, food safety as well as flexibility to alleviate the varying needs of the market and in the same time opening the door towards promotion of efficiency and in resilience of operations. As stressed by Moses and Anandharamakrishnan, the mentioned technologies can serve as an essential factor that helps transform the conventional systems into very elegant and high-level automated and optimised systems in respect to the Industry 5.0 paradigm (Moses et al., 2024).

6.1.1 Optional saviors of food production industry: Artificial Intelligence (AI) and Internet of Things (IoT):

Monitoring in the food manufacturing supply chain can be performed in real-time and data-based decisions taken with the combination of AI and IoT. Making use of the available resources, maximising the crop yield throughout the board, and making the production process more efficient, precision farming powered by AI employs the output of the IoT-enabled sensors, as Misra et al., 2020 and Ding et al., 2023 say. IoT devices and AI algorithms in predictive maintenance can be used to prevent wastage in the food industry to a substantial extent, as well as avoiding equipment breakdown that causes a downtime in the equipment to be lower. The resource optimizing resource-managing and carbon minimizing routing and scheduling in artificial intelligence in logistics and supply chain management is a source of pride because it contributes to sustainability of food manufacturing in general (Hassoun et al., 2023). The new study also confirms how implementation of blockchain along with AI and IoT would also contribute to the end-to-end traceability within the supply chain, enhancing the transparency, and contributing to the limited occurrence of fraud (Ellahi et al., 2023). Hopefully, this synergy is likely to counter the problem of scalability and interoperability and consequently experience an efficient and smooth supply chain network.

6.1.2 RAS: drive: efficiency and effective use of time:

With the brining of RAS, the total impacts of AI in food production are also enhanced. According to Kumar et al., 2011 the robotic systems also enhance consistency and lower the cost of labor besides automating these functions. AI integration will enable the company to control the production processes accurately, offering uniform quality and effectiveness. As an example, a machine vision can be used as the robotic sorting system to detect defective products and dispose of them with a high level of specificity and efficiency, ensuring food safety standards. Manufacturing plants are also characterized by the use of autonomous vehicles that make better use of the workflow by streamlining the handling of materials and cutting operational costs (Hassoun et al., 2023). As well, cooperative robots, or cobots, have increased under consideration to boost person-robot interaction on manufacturing floors by building a secure and even more productive working environment.

6.1.3 Digital twins: optimisation-in-process through the processes of virtualization:

Application of DT technology is an excellent move in the optimization of food manufacturing process. DT allows simulating, monitoring, and optimizing the production processes in real-time through the development of virtual copies of physical processes. The Grewal et al., 2019 argue that DT streamlines resource provision, reduces by-product, and also maximizes energy use, which presents aspect of sustainability. Its flexibility and scalability make DT an effective instrument in managing the needs of food industry, which change over time. The further development of DT should integrate ML models in predictive analytics so that manufacturers could predict the emergence of bottlenecks and prevent them through proactive strategies. Such practice would increase resilience and sustainability of the food manufacturing ecosystem (Moses et al., 2024).

6.1.4 Industry 5.0 food manufacturing industry:

The future of industries is industry 5.0 which is a continuation of the digital revolution of industry 4.0, the human element, environmental-friendliness and flexibility with regard to manufacturing. Unlike Industry 4.0, Industry 5.0 focuses on collaborative intelligence, where both human beings and advanced technologies cooperate and apply their wisdom to enhance efficiency in the production process, the environment, true sustainability and the general resilience (Hassoun et al., 2023, Nahavandi et al., 2019). Among the significant Industry 5.0 principles applicable to food production, there must be the following principles:

- Human-AI cooperation: compared to the fully automated Industry 4.0, considered to be based on the idea of human-AI collaboration, Industry 5.0 is characterized by human-AI relationship, whether AI assists human employees in their valuable decision-making, but it also provides flexibility and conformity throughout the process of manufacturing (119).
- Sustainable manufacturing: the shift towards the world of the circular economy also picks up additional steam under Industry 5.0, where the process of manufacturing is underpinned by the use of AI, IoT and robotics in an effort to reduce wastage, reusing by-products, and consumption of energy (Rame et al., 2024).
- The resilience and adaptive structures: industry 5.0 generates resilience by having the capability of smart manufacturing system to act quickly in the event of a supply chain process disruption that will guarantee the sustainability and stability of food production process (Van Erp et al., 2024).

Table 1. Food manufacture: major innovations and prospects.

Technology	Major Innovations	Future Prospests and Advancements
Sustainable AI	- Real-time monitoring and predictive analytics (Misra et al., 2020) - Precision farming for optimized resource utilization (Ding et al., 2023) - Enhanced waste management (Singagerda et al., 2025)	- Democratizing AI adoption across all farm scales (**Koebe, 2025**) - Integration with Industry 5.0 for human-centric automation (Redchuk et al., 2023)
Internet of Things (IoT)	- Sensor-driven decision-making for accurate field interventions (Ding et al., 2023) - Predictive maintenance reducing equipment downtime (Hassoun et al., 2023).	- Greater scalability and cross -device interoperability (Ellahi et al., 2023) - Blockchain integration for transparent traceability (Ellahi et al., 2023)
Robot and Autonomous systems (RAS)	- Automated sorting and defect detection for quality control **(Kumar et al., 2011)** - Use of collaborative robots for safer, efficient workplaces (Hassoun et al., 2023).	- Smarter human-robot interaction interfaces (Hassoun et al., 2023). - Increased efficiency in logistics and material handling (Hassoun et al., 2023).
Digital Twins (DT)	- Creation of virtual replicas for real-time simulation and process control (Grewal et al., 2019) - Optimization of resource management in production (**Moses et al., 2024**)	- Predictive analytics through machine learning integration (**Moses et al., 2024**) - Responsive systems for dynamic production demands (Grewal et al., 2019)
AI-driven Innovations	- Personalized nutrition services using machine learning (Viejo et al., 2019) - AI-based quality control and assurance mechanisms (Zuo et al., 2024)	- Improved regulatory adherence using AI tools (Zuo et al., 2024) - Customized food production addressing individual health needs (Viejo et al., 2019)
Circular Economy and Renewable Energy	- Reuse of surplus ingredients and agro-industrial by-products (Duckett et al., 2018) - Implementation of solar and other renewable energy sources (Singagerda et al., 2025)	- Stronger alignment with climate and sustainability targets (**Koebe, 2025**) - Enhanced energy efficiency across all stages of the supply chain (Redchuk et al., 2023)

The advantages of using the principles of the Industry 5.0 concept in the food manufacturing process are numerous, as they include the intensification of AI-aided predictive maintenance, the creation of the concept of sustainable food production and the implementation of robotics and personalized nutrition technology in favor of targeted food production. Through the application of these technologies, the food sector will be able to move to a more sustainable, consumer-friendly, and ecologically responsible output system (Hassoun et al., 2023).

6.1.5 A development of AI solutions: individual nutrition and improvement of quality assurance:

Testing Artificial Intelligence is fueling new developments within the areas of personal nutrition and the possession of food production. Creation of individual nutritional demands and preferences helps to customize food products through ML models as highlighted by Viejo et al. (Viejo et al., 2019). By utilizing AI, additive manufacturing allows creating customized food products with nutritional improvements, filling in the gap between being healthy and convenient (Yu et al., 2024). Artificial intelligence application in food safety Food safety A third developing area of AI application is in quality control systems involving computer vision and DL classes to monitor contaminants, freshness, and shelves life. The discussed technologies enhance the quality and effectiveness of the quality assurance procedures to a great extent, and it can now be included in the strict regulations (Viejo et al., 2019, Zuo et al., 2024).

6.1.6 Conquering the obstacles and writing your future:

Even though AI technologies offer great potential, they are also being undermined by several barriers, which limit their usage in the food manufacturing industry. Implementation costs, reliance on data and ethical issues of privacy and bias continue to be a major impediment (Agrawal et al., 2023). The only way to overcome the mentioned challenges is to apply a holistic approach in which governments, academia, and industry players will collaborate and implement cost-efficient solutions and build effective regulations (Czaja et al., 2025). The enhancement of interpretability and transparency of AI models should be the future research priorities to establish the stakeholder trust. Moreover, upskilling of workforce and interdisciplinary projects will be required to ultimately take advantage of these technologies. The prospect of the road implies that AI will be deployed as the mode of balancing innovation, sustainability, and ethical considerations and implications of food production (Hassoun et al., 2023, Moses et al., 2024).

6.2 The sustainable food production into green future: AI

It is the use of food production AI, big data, and IoT, which are encouraging radical progress in the field of sustainability. The technologies are more efficient in operations, utilization of resources and minimize wastages. Indeed, as noted by Misra et al., 2020, the possibility to have real-time monitoring and insights into the prediction can assist the manufacturers to enhance the processes, minimize the losses, and address the environmental issues as required by the Agenda 2030. As Koebe, 2025 point out, the idea behind the extensive implementation of digital technologies, including AI, in fulfilling the objectives of SDGs, drafted by the United Nations, can be used. The scalability of the AI-powered model and the digital channels concepts facilitates the sustainable activity and allows global versatility, even the enterprises that are emerging in the developing market. Artificial Intelligence applied 3D food printing is an outstanding innovation in food production and would create less waste as well as the possibility of creating a finer volume of food in customized samples that consumers enjoy. The use of surplus ingredients and by-products in the formulation would be a proactive step to make circular economy in food production a true reality, since it will largely decrease food waste (Duckett et al., 2018). The illustration of the sustainable food manufacture big picture (Figure 1) emphasizes the three necessary components, such as the process optimization, seamless implementation of digital technologies, and embracement of solutions that involve AI and the Internet of Things. This framework supports the fact that blending technological innovation and efficiency of the processes can achieve long-term sustainability objectives.

Figure 1. Sustainable practices in food manufacturing

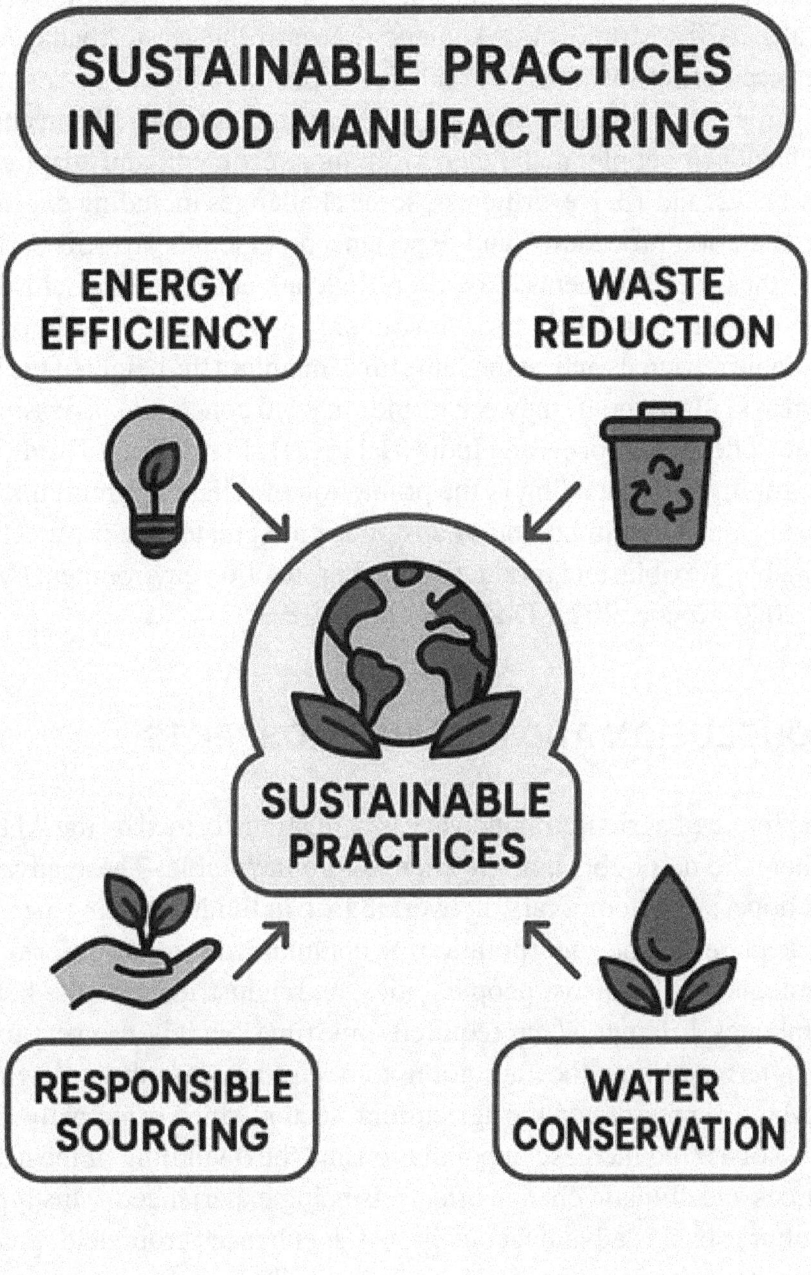

The energy management systems driven by AI also contribute to the concept of sustainability as they manage the usage of energy throughout the food chain, carbon footprints, and decrease costs of operations (Misra et al., 2020). Moreover, AI integration into the waste management systems can bring sustainable urban development, which is discussed by Singagerda et al. (Singagerda et al., 2025), who claim that AI can change the behavior of people, make their work more efficient, and adapt to the SDGs 11, 12, and 13. Nevertheless, some challenges including expensive implementation procedures and experienced personals are still on hand despite these developments. However, Redchuk et al., 2023 disclose that among the solutions to help to achieve energy efficiency and optimization of available resources and, at the same time, maintain the height of the level of human skills in the framework of Industry 5.0 concept, it is possible to use low-code AI platforms and Industrial IoT (IIoT) solutions. Eventually, the possibility of AI leading to the promotion of reliability, minimization of wastage, and maximization of customisation creates the prospect of a sustainable, flexible, and productive food production environment (Misra et al., 2020, Koebe, 2025, Duckett et al., 2018).

7. CONCLUSION AND FUTURE PROSPECTS

The future of agricultural industry is being transformed by the AI trend and there are no doubts that the changes are inevitable. These advances bring hope in contemporary networking of individuals using personal cellular phones. The incorporation of communication and information technologies in the life of people, cities, and industries with the help of smartphones, IoT, and AI has required some time, but it is also very promising in terms of how they are going to be used in case the following is true. Massive pressure on the agricultural sector, which principally arises because of a rapid increase in population and the dwindling of the natural resources and climatic change effects, is being experienced. This type of agriculture stage is advantageous in that, it enhances crop yield, guarantees optimal management choices, avoids pollution of the environment since the application of chemicals is minimized, and conserve resources, such as, electricity, water and Climate-smart agriculture embraces climate resilience through use of carbon on soil, decrease in emission of greenhouse gases, and an increase in farm and farmer productivity and revenue to facilitate their income and food security in a fast changing climate. The combination of the strength of IoT and precision farming leads to a network of advanced tools of farming. Precision farming offers the solution to the issue of agriculture by borrowing ideas of the Internet of Things (IoT) technologies. Speedy gathering of farm data enables farmers to devise plans, which are specific to their farm and the climatic situation of that year. Application of precision farming to aspects such as farming irrigation, farmers livestock, and equipments follow are paramount to the enhancement of improvement in productivity and performance. Through AI technologies the farmers can examine the state of the soil and whether there is any relative information with the aim of making the operations more efficient. More than that, the technologies permit real-time observation of the performance of interconnected devices to avoid overflowing amounts of water and nutrients.

Farmers can monitor trends and forecast what weather next will bring and what crop yield they will expect by utilizing the AI technologies. Internet of Things (IoT) applications in agricultural enable farmers to maintain the crop quality and soil fertility, which will lead to improved yield and quality. Using the data, technological advancements are harnessed which empower better decisions for agriculture. Through the use of sensor information, the IoT machinery provides timely status reports on crops to regulate farming operations. Using predictive analytics, farmers will be able to collect data for better decision making regarding harvesting. The development of agriculture is facilitated if the process of scientifically proven planting, the product of automation machinery, and improvement of production processes are implemented. The smart technology can accelerate the movement of the region into advanced practices of the AI. In precision farming, photos and sensors are used to gather information that is used in observation of real time crop, soil and air closely so that changes to the farming activities can be easily detected and addressed by the farmer. Device for crop management is a part of precision farming strategies and is identified as one of the key IoT devices in the agriculture.

Data analytics for prediction is critical to the precision agriculture practices. Although these IoT and smart sensors offer a lot of updated data, the data analytics aids farmers to walk around the data and come up with critical insights like the when to harvest, likelihood of epidemic and plague attack, and actual output volumes. These tools are integral in helping streamline and forecast the outcomes of farming heavily. These instruments, like those of weather stations will be erected in the field itself and will obtain essential information one that pertains to the cultivation of crops namely, temperature, rainfall, the water potential of the leaf and the world crop health. Other than the collection of data regarding the weather, weather stations are able to set controls to fit predetermined standards even without the direction of a human in the future. A large mobile phone connects the drivers of small machinery to the crop management equipment, which is essential to precision farming and one of the most exemplifying components of IoT technology applied in the agricultural sector. Such devices will be put into the agricultural fields the same way as weather stations are to monitor individual variables through which crops are grown for example temperature, rainfall, leaf water potential and crop health levels among others.

8. REFERENCES

Agrawal, K., & Kumar, N. (2023). Artificial Intelligence Innovations: Inception of new horizons in food processing sector. In: *2023 IEEE Silchar Subsection Conference (SILCON)*. (p. 1–8. doi: DOI: 10.1109/SILCON59133.2023.10404183

Aker, J. C., Ghosh, I., & Burrell, J. (2016). The promise (and pitfalls) of ICT for agriculture initiatives. *Agricultural Economics*, *47*(S1), 35–48.

Azadi, H., Moghaddam, S. M., Burkart, S., Mahmoudi, H., Van Passel, S., Kurban, A., & Lopez-Carr, D. (2021). Rethinking resilient agriculture: From climate-smart agriculture to vulnerable-smart agriculture. *Journal of Cleaner Production*, *319*, 128602.

Balasundram, S. K., Shamshiri, R. R., Sridhara, S., & Rizan, N. (2023). The role of digital agriculture in mitigating climate change and ensuring food security: An overview. *Sustainability*, *15*(6), 5325.

Behera, D., & Modak, S. (2022). Attitude and opinion of the farmers towards regulated market, a case of Balasore District. *Indian Research Journal of Extension Education*, *22*(2), 169–172.

Bhattacharjee, S., & Raj, S. (2016). *Social media: Shaping the future of agricultural extension and advisory services. GFRAS interest group on ICT4RAS discussion paper*. GFRAS.

Bhattacharyya, T., Wani, S. P., & Tiwary, P. (2021). Empowerment of stakeholders for scaling-up: Digital technologies for agricultural extension. In *Scaling-up solutions for farmers* (pp. 121–147). Springer.

Cole, S., & Sharma, G. (2017). The promise and challenges of implementing ICT in Indian agriculture. In *India Policy Forum* (pp. 11-12).

Crop, X. (2025). CropX agronomic farm management system. CropX. Retrieved from https:// cropx.com/

Czaja, T. P., & Engelsen, S. B. (2025). Why nothing beats NIRS technology: The green analyticalchoice for the future sustainable food production. *Spectrochimica Acta. Part A: Molecular and Biomolecular Spectroscopy*, *325*, 125028. DOI: 10.1016/j.saa.2024.125028 PMID: 39217952

Das, B. (2014). ICTs Adoption for Accessing Agricultural Information: Evidence from Indian Agriculture. *Agricultural Economics Research Review*, *27*(2), 199–208.

Ding, H., Tian, J., Yu, W., Wilson, D. I., Young, B. R., & Cui, X.. (2023). The applicationof artificial intelligence and big data in the food industry. *Foods*, *12*, 4511. DOI: 10.3390/foods12244511 PMID: 38137314

Duckett, T., Pearson, S., Blackmore, S., & Grieve, B. (2018). *Agricultural Robotics: The Future of Robotic Agriculture*. EPSRC UK-RAS Network., DOI: 10.31256/WP2018.2

Ellahi, R. M., Wood, L. C., & Bekhit, A. E. D. A. (2023). Blockchain-based frameworks for foodtraceability: A systematic review. *Foods*, *12*, 3026. DOI: 10.3390/foods12163026 PMID: 37628025

. M.A. Ferrag, L. Shu, O. Friha, X. Yang, Cyber security intrusion detection for agriculture 4.0: machine learning-based solutions, datasets, and future directions, IEEE/CAA Journal of Automatica Sinica 9 (3) (2021) 407–436.

FieldRoutes. (2025). AI in pest control: A game-changer for the industry. Retrieved from https://www.fieldroutes.com/blog/pest-control-ai

Fraser, E. D., & Campbell, M. (2019). Agriculture 5.0: Reconciling production with planetary health. *One Earth*, *1*(3), 278–280.

Gagliardi, G., Lupia, M., Cario, G., Cicchello Gaccio, F., D'Angelo, V., Cosma, A. I. M., & Casavola, A. (2021). An internet of things solution for smart agriculture. *Agronomy (Basel)*, *11*(11), 2140.

Goyal, A., & Mishra, D. (2016). Will digital technologies transform agriculture in developing countries? *Agricultural Economics*, *47*(S1), 21–33.

Grewal, D., Hulland, J., Kopalle, P. K., & Karahanna, E. (2019). The future of technologyand marketing: A multidisciplinary perspective. *Journal of the Academy of Marketing Science*, *48*, 1–8. DOI: 10.1007/s11747-019-00711-4

Gryshova, I., Balian, A., Antonik, I., Miniailo, V., Nehodenko, V., & Nyzhnychenko, Y. (2024). Artificial intelligence in climate smart in agricultural: Toward a sustainable farming future. Access to science, business, innovation in the digital economy. *ACCESS Press*, *5*(1), 125–140.

Guo, Y., & Zhang, C. (2021). Recent Advances in Large Margin Learning. *IEEE Transactions on Pattern Analysis and Machine Intelligence*. PMID: 34161238

Hassoun, A., Aït-Kaddour, A., Dankar, I., Safarov, J., Ozogul, F., & Sultanova, S. (2023). The significance of industry 4.0 technologies in enhancing various unit operations applied in the food sector: Focus on food drying. *Food and Bioprocess Technology*, *18*, 109–128. DOI: 10.1007/s11947-024-03465-2

Huyer, S. (2021). Gender-smart agriculture: An agenda for gender and socially inclusive climate-resilient agriculture. *Journal of Cleaner Production*, *328*, 129720.

IBM. (2025). Watson Decision Platform for Agriculture. Retrieved from https://www.ibm.com

Koebe, P. (2025). How digital technologies and AI contribute to achieving the health-related SDGs. *International Journal of Information Management Data Insights*, *5*, 100298. DOI: 10.1016/j.jjimei.2024.100298

Kumar, A., & Konar, A. (2011). *Dyeing of Textiles with Natural Dyes*. InTech., DOI: 10.5772/21341

Liu, W., Shao, X. F., Wu, C. H., & Qiao, P. (2021). A systematic literature review on applications of information and communication technologies and blockchain technologies for precision agriculture development. *Journal of Cleaner Production*, *298*, 126763.

Misra, N., Dixit, Y., Al-Mallahi, A., Bhullar, M. S., Upadhyay, R., & Martynenko, A. (2020). IoT, bigdata, and artificial intelligence in agriculture and food industry. *IEEE Internet of Things Journal*, *9*, 6305–6324. DOI: 10.1109/JIOT.2020.2998584

Moses, J. A., & Anandharamakrishnan, C. (2024). Emerging Technologies for the Food Industry: Vol. 3. *ICT Applications and Future Trends in Food Processing*. Apple Academic Press., DOI: 10.1201/9781003413660

Nahavandi, S. (2019). Industry 5.0—A human-centric solution. *Sustainability*, *11*, 4371. DOI: 10.3390/su11164371

Palanisamy, A., & Bharadwaj, N. (2017). Utilization of Information Disseminated through Mobile Telephones by Farmers in Tamil Nadu. *Journal of Extension Education*, *29*(3), 5902–5909. DOI: 10.26725/jee.2017.3.29.5902-5909

Plantix. (2025). #1 free app for crop diagnosis and treatments. Retrieved from https:// plantix.net/en/

Rame, R., Purwanto, P., & Sudarno, S. (2024). Industry 5.0 and sustainability: An overview ofemerging trends and challenges for a green future. *Innov Green Dev*, *3*, 100173. DOI: 10.1016/j.igd.2024.100173

Redchuk, A., Walas Mateo, F., Pascal, G., & Tornillo, J. E. (2023). Adoption case of IIoT and machine learning to improve energy consumption at a process manufacturing firm, under industry 50 model. *Big Data and Cognitive Computing*, *7*, 42. DOI: 10.3390/bdcc7010042

Rolandi, S., Brunori, G., Bacco, M., & Scotti, I. (2021). The digitalization of agriculture and rural areas: Towards a taxonomy of the impacts. *Sustainability*, *13*(9), 5172.

Saravanan, R. (2012). ICTs for agricultural extension in India: policy implications for developing countries. In Proc. of 8th Asian Conference for Information Technology in Agriculture, (AFITA) (pp. 1-11).

Singagerda, F. S., Dewi, D. A., Trisnawati, S., Septarina, L., & Dhika, M. R. (2025). Towards a circular economy: Integration of AI in waste management for sustainable urban growth. *J Lifestyle SDG's Rev*, *5*, e02642. DOI: 10.47172/2965-730X.SDGsReview.v5.n01.pe02642

Smith, J., Brown, T., & Wilson, A. (2023). Sustainable resource management and AI applications in climate-smart agriculture. *Journal of Agricultural Innovation and Sustainability*, *12*(3), 45–57. https://doi.org/10.xxxx/jais.2023.12345

Spanaki, K., Karafili, E., & Despoudi, S. (2021). AI applications of data sharing in agriculture 4.0: A framework for role-based data access control. *International Journal of Information Management*, *59*, 102350.

Sponchioni, G., Vezzoni, M., Bacchetti, A., Pavesi, M., & Renga, F. M. (2019). The 4.0 revolution in agriculture: a multi-perspective definition. In *Summer School F* (pp. 143–149). Turco-Industrial Systems Engineering.

Symeonaki, E., Arvanitis, K. G., Loukatos, D., & Piromalis, D. (2021). Enabling IoT wireless technologies in sustainable livestock farming toward agriculture 4.0. In *IoT- based Intelligent Modelling for Environmental and Ecological Engineering* (pp. 213–232). Springer.

Tao, W., Zhao, L., Wang, G., & Liang, R. (2021). Review of the internet of things communication technologies in smart agriculture and challenges. *Computers and Electronics in Agriculture*, *189*, 106352.

. Van Erp T, Carvalho NGP, Gerolamo MC, Gonçalves R, Rytter NGM,Gladysz B. (2024). Industry 5.0: a new strategy framework for sustainability management and beyond. *J Clean Prod.* (2024) 461:142271. doi: . 142271DOI: 10.1016/j.jclepro.2024

Viejo, C. G., Torrico, D. D., Dunshea, F. R., & Fuentes, S. (2019). Emerging technologies basedon artificial intelligence to assess the quality and consumer preference of beverages. *Beverages*, *5*, 62. DOI: 10.3390/beverages5040062

Weres, J., Mueller, W., Kozłowski, R., & Kluza, T. Advanced technologies in developing Web-based decision support systems for agriculture, in: *Proc. Of the CIGR XVIIth World Congress and the 8th World Congress on Computers in Agriculture*, 2010.

Xu, Z., Lv, Z., Li, J., Sun, H., & Sheng, Z. (2022). A Novel Perspective on Travel Demand Prediction Considering Natural Environmental and Socioeconomic Factors. *IEEE Intelligent Transportation Systems Magazine*.

. X. Yang, L. Shu, J. Chen, M.A. Ferrag, J. Wu, E. Nurellari, K. Huang, A survey on smart agriculture: development modes, technologies, and security and privacy challenges, IEEE/CAA Journal of Automatica Sinica 8 (2) (2021) 273–302.

Yu, Q., Zhang, M., Mujumdar, A. S., & Li, J. (2024). AI-based additive manufacturing for futurefood: Potential applications, challenges and possible solutions. *Innovative Food Science & Emerging Technologies*, *92*, 103599. DOI: 10.1016/j.ifset.2024.103599

Zuo, M., Ji, H., Su, L., Zhang, Y., Yan, W., & Zhang, Q.. (2024). Advancements andmodern applications of artificial intelligence in ensuring food safety. *J Chin Inst Food Sci Technol*, *24*, 1–13.

Conclusion

The rise of AI Driven Smart Agriculture is a result of many challenges in agriculture presented by ongoing increase of the unpredictable and very powerful natural phenomena worldwide. The book presents a selected collection of peer reviewed contributions addressing technological, economic, political and ethical issues concerning the research, development and innovation in the field of Smart Agriculture and Health Food Production.

As environmental dramatic changes continue to disrupt traditional agricultural practices, the integration of smart technologies has become essential for sustainable food systems. The authors valuable contributions present the cutting-edge innovations are transforming the way people cultivate and harvest food. The studies show a clear path to achieve global food security while safeguarding the health of both people and the planet.

The rise of Applied AI in Smart Agriculture will drive the innovation and transforming the next generation agriculture across the globe, and its long rooted in traditional methods are no exception. The rise of transformative powerful catalyst for change will drive a new era in food production that emphasizes efficiency, sustainability, and precision.

The book opens new horizons in Climate-Resistant Smart Agriculture for Healthy Food Production, while exploring new developments in the field of ultra-smart agriculture to face the climate change. This book presents the most current technological and business solutions that will drive future smart agriculture and healthy food production for the benefit of mankind. The book promotes creation of national and global multidisciplinary research teams to seek best solutions for agriculture, climate change, and smart technology today and tomorrow.

The book is an excellent resource for scholars, policymakers and decision makers, environmental scientists, agricultural professionals, and general public.

Prof. Eduard Babulak

2025-05-08

Compilation of References

Aayog, N. I. T. I. (2023). *Efficiency of Micro-Irrigation in Economizing Water Use in India: Learning from Potential and Under Explored States*. Government of India. https://www.niti.gov.in/sites/default/files/2023-03/Efficiency%20of%20Micro-Irrigation%20in%20economizing%20water%20use%20in%20India.pdf

Abdullah, H. M., Mohana, N. T., Khan, B. M., Ahmed, S. M., Hossain, M., Islam, K. S., Redoy, M. H., Ferdush, J., Bhuiyan, M., Hossain, M. M., & Ahamed, T. (2023). Present and future scopes and challenges of plant pest and disease (P&D) monitoring: Remote sensing, image processing, and artificial intelligence perspectives. *Remote Sensing Applications: Society and Environment*, *32*, 100996. DOI: 10.1016/j.rsase.2023.100996

Adewusi, A. O., Asuzu, O. F., Olorunsogo, T., Iwuanyanwu, C., Adaga, E., & Daraojimba, D. O. (2024). AI in precision agriculture: A review of technologies for sustainable farming practices. *World Journal of Advanced Research and Reviews*, *21*(1), 2276–2285.

Agrawal, K., & Kumar, N. (2023). Artificial Intelligence Innovations: Inception of new horizons in food processing sector. In: *2023 IEEE Silchar Subsection Conference (SILCON)*. (p. 1–8. doi: DOI: 10.1109/SILCON59133.2023.10404183

Agus, F., Surmaini, E., & Lina, H. (2023). Strategies for Indonesia's Agricultural Climate Change Adaptation and Mitigation. *Jurnal Sumberdaya Lahan*, *16*(2), 67. DOI: 10.21082/jsdl.v16n2.2022.67-81

Ahamed, T. (2024). *IoT and AI in Agriculture: Smart Automation Systems for Increasing Agricultural Productivity to Achieve SDGs and Society 5.0.* Springer Nature.

Ahuja, S., Manjunath, H. R., Alam, I., & Rastogi, A. (2024). The Role of AI in Modern Farming: Precise Pest Management and Optimal Water Use. *International Journal of Chemical and Biochemical Science*, *25*(13), 249–255.

Aijaz, N., Lan, H., Raza, T., Yaqub, M., Iqbal, R., & Pathan, M. S. (2025). Artificial intelligence in agriculture: Advancing crop productivity and sustainability. *Journal of Agriculture and Food Research*, •••, 101762.

Ajzen, I. (1991). The theory of planned behavior. *Organizational Behavior and Human Decision Processes*, *50*(2), 179–211. DOI: 10.1016/0749-5978(91)90020-T

Akanbi, M. B., Banjoko, I. K., Adedotun, K. J., & Raji, A. K. (2024). AI-Powered Smart Irrigation Systems and Solar Energy Integration: A Sustainable Approach to Enhancing Agricultural Productivity in Nigeria. *Journal of Renewable Agricultural Technology Research.*

Aker, J. C., Ghosh, I., & Burrell, J. (2016). The promise (and pitfalls) of ICT for agriculture initiatives. *Agricultural Economics*, *47*(S1), 35–48.

Ali, A., Tajamul, H., & Azlan, Z. (2025). Smart irrigation technologies and prospects for enhancing water use efficiency for sustainable agriculture. *AgriEngineering*, *7*(4), 106.

Ali, G., Mijwil, M. M., Buruga, B. A., Abotaleb, M., & Adamopoulos, I. (2024). A survey on artificial intelligence in cybersecurity for smart agriculture: State-of-the-art, cyber threats, artificial intelligence applications, and ethical concerns. *Mesopotamian Journal of Computer Science*, *2024*, 53–103.

Altieri, M. A., & Nicholls, C. I. (2017). The adaptation and mitigation potential of traditional agriculture in a changing climate. *Climatic Change*, *140*, 33–45.

Amulothu, D. V. R. T., Rodge, R. R., Hasan, W., & Gupta, S. (2024). Machine Learning for Pest and Disease Detection in Crops. In *Agriculture 4.0* (pp. 111–132). CRC Press.

Antonelli, M., & Donelli, D. (2023, October). Precision Nutrition and Artificial Intelligence Mobile Apps: A Narrative Review. In *Biology and Life Sciences Forum* (Vol. 29, No. 1, p. 25). MDPI.

Appendini, K., & García-Barrios, R. (2003). *Food security and quality in Mexico: A report to FAO. Food and Agriculture Organization.* FAO.

Ashoka, P., Devi, B. R., Sharma, N., Behera, M., Gautam, A., Jha, A., & Sinha, G. (2024). Artificial Intelligence in Water Management for Sustainable Farming: A Review. *Journal of Scientific Research and Reports, 30*(6), 511–525.

Asian Development Bank. (2022). *Rural Development and Inclusive Growth in Asia.* https://www.adb.org/publications

Assimakopoulos, F., Vassilakis, C., Margaris, D., Kotis, K., & Spiliotopoulos, D. (2024). The implementation of "smart" technologies in the agricultural sector: A review. *Information (Basel), 15*(8), 466.

Assimakopoulos, F., Vassilakis, C., Margaris, D., Kotis, K., & Spiliotopoulos, D. (2025). AI and related technologies in the fields of smart agriculture: A review. *Information (Basel), 16*(2), 100.

Atkinson, A., & Messy, F. A. (2012). Measuring financial literacy: Results of the OECD/International Network on Financial Education (INFE) pilot study. *OECD Working Papers on Finance, Insurance and Private Pensions,* No. 15. OECD Publishing. https://doi.org/DOI: 10.1787/5k9csfs90fr4-en

Ayeni, O., & Olagoke-Komolafe, O. E. (2024). Advancing food safety standards through technology integration and policy development. 2(1), 035–046. https://doi.org/DOI: 10.57219/crrms.2024.2.1.0039

Azadi, H., Moghaddam, S. M., Burkart, S., Mahmoudi, H., Van Passel, S., Kurban, A., & Lopez-Carr, D. (2021). Rethinking resilient agriculture: From climate-smart agriculture to vulnerable-smart agriculture. *Journal of Cleaner Production, 319*, 128602.

Azad, T., & Ahmed, S. (2016). Common milk adulteration and their detection techniques. *International Journal of Food Contamination.* Advance online publication. DOI: 10.1186/s40550-016-0045-3

Aziz, D., Rafiq, S., Saini, P., Ahad, I., Gonal, B., Rehman, S. A., Rashid, S., Saini, P., Rohela, G. K., Aalum, K., & Nabila Iliya, M. (2025). Remote sensing and artificial intelligence: Revolutionizing pest management in agriculture. *Frontiers in Sustainable Food Systems, 9*, 1551460.

Balasundram, S. K., Shamshiri, R. R., Sridhara, S., & Rizan, N. (2023). The role of digital agriculture in mitigating climate change and ensuring food security: An overview. *Sustainability, 15*(6), 5325.

Bano, A., Ali, M., Gupta, A., Pathak, N., & Hasan, W. (2021). *Climate Smart Agriculture and Hi-Tech Farming.*

Baruah, R. (2024). Restoring Soil Health of Degraded Lands Through Eco-Friendly Nutrient Management Practices in Regenerative Agriculture. In *Regenerative Agriculture for Sustainable Food Systems* (pp. 237–269). Springer Nature Singapore.

Bates, D. W., Auerbach, A., Schulam, P., Wright, A., & Saria, S. (2020). Reporting and implementing interventions involving machine learning and artificial intelligence. *Annals of Internal Medicine, 172*(11, Supplement), S137–S144. PMID: 32479180

Behera, D., & Modak, S. (2022). Attitude and opinion of the farmers towards regulated market, a case of Balasore District. *Indian Research Journal of Extension Education, 22*(2), 169–172.

Beniwal, V., Rana, K., & Singh, S. (1999). Detection of commonly used adulterants in foodstuffs and consumers' awareness about them. *Journal of Food Science and Technology, 36*(3), 273–275.

Bhat, S. A., & Huang, N. F. (2021). Big data and ai revolution in precision agriculture: Survey and challenges. *IEEE Access : Practical Innovations, Open Solutions*, *9*, 110209–110222.

Bhattacharjee, S., & Raj, S. (2016). *Social media: Shaping the future of agricultural extension and advisory services. GFRAS interest group on ICT4RAS discussion paper*. GFRAS.

Bhattacharyya, T., Wani, S. P., & Tiwary, P. (2021). Empowerment of stakeholders for scaling-up: Digital technologies for agricultural extension. In *Scaling-up solutions for farmers* (pp. 121–147). Springer.

Bhuiyan, M. M. R., Noman, I. R., Aziz, M. M., Rahaman, M. M., Islam, M. R., Manik, M. M. T. G., & Das, K. (2025). Transformation of Plant Breeding Using Data Analytics and Information Technology: Innovations, Applications, and Prospective Directions. *Frontiers in Bioscience-Elite*, *17*(1), 27936.

Bogadi, N. P., Banović, M., & Babić, I. (2016). *Food defence system in food industry: perspective of the EU countries*. Journal für Verbraucherschutz und Lebensmittelsicherheit Journal of Consumer Protection and Food Safety., DOI: 10.1007/s00003-016-1022-8

Bouzembrak, Y., Klüche, M., Gavai, A., & Marvin, H. J. P. (2019). Internet of Things in food safety: Literature review and a bibliometric analysis. *Trends in Food Science & Technology*, *94*, 54–64. DOI: 10.1016/j.tifs.2019.11.002

Braimoh, A. K., Emenanjo, I., Rawlins, M. A., Heumesser, C., & Zhao, Y. (2016). Climate-smart agriculture indicators. 1–126. https://documents.worldbank.org/curated/en/187151469504088937/Climate-smart-agriculture-indicators

Brundtland Commission. (1987). *Our common future: Report of the World Commission on Environment and Development*. Oxford University Press.

Budhwar, M., Kumar, A., & Singh, K. (2023). IoT adoption in food processing industry: An integrated Fuzzy-DEMATEL and Fuzzy-TOPSIS approach. *Journal of Agriculture and Food Research*, *14*, 100776. DOI: 10.1016/j.jafr.2023.100776

Bulut, H., & Arslan, S. (2022). Smart Agriculture: IoT and Artificial Intelligence Applications in Modern Farming. *Agricultural Technology Journal*, *34*(2), 101–115.

Campera, M., Budiadi, B., Adinda, E., Ahmad, N., Balestri, M., Hedger, K., & Nekaris, K. A. I. (2021). Fostering a wildlife-friendly program for sustainable coffee farming: The case of small-holder farmers in Indonesia. *Land (Basel)*, *10*(2), 121.

Cartolano, A., Cuzzocrea, A., & Pilato, G. (2024). Analyzing and assessing explainable AI models for smart agriculture environments. *Multimedia Tools and Applications*, *83*(12), 37225–37246.

Casillas, C. E., & Kammen, D. M. (2010). The energy-poverty-climate nexus. *Science*, *330*(6008), 1181–1182. DOI: 10.1126/science.1197412 PMID: 21109654

Challinor, A. J., Watson, J., Lobell, D. B., Howden, S. M., Smith, D. R., & Chhetri, N. (2014). A meta-analysis of crop yield under climate change and adaptation. *Nature Climate Change*, *4*(4), 287–291.

Chang, C.-C., & Hsu, S.-C. (2021). Smart agriculture policy in Taiwan: Opportunities and challenges in climate-resilient farming. *Journal of Agricultural Policy and Development*, *34*(2), 115–130. DOI: 10.1016/j.japd.2021.02.005

Chellaram, C., Rajamalar, C. G., & Chandrika, M. (2014). Nanotechnology applications in food industry. [). Elsevier B.V.]. *APCBEE Procedia*, *8*, 109–113.

Chen, C. F., & Chen, M. C. (2023). Integrating AI and IoT in Taiwan's smart agriculture: Enhancing resilience to climate change. *Journal of Environmental Management*, *330*, 117180. DOI: 10.1016/j.jenvman.2022.117180

Chen, H., & Volpe, R. P. (2019). An analysis of personal financial literacy among college students. *Financial Services Review, 7*(2), 107–128.

Chen, L. Y., Lee, Y. C., & Huang, S. T. (2020). Regional collaboration in climate-smart agriculture: A pathway for Asia-Pacific food resilience. *Sustainability, 12*(17), 6892. DOI: 10.3390/su12176892

Chib, S., Mehta, A. K., Selvakumar, P., Saravanan, A., Mishra, B. R., & Manjunath, T. C. (2025). Mental Health Challenges and Solutions in High-Pressure Work Environments. In Harriott, S. (Ed.), *Innovative Approaches to Managing Conflict and Change in Diverse Work Environments* (pp. 247–272). IGI Global Scientific Publishing.

Chou, Y. C., Wu, P. J., & Lin, C. Y. (2022). Digital transformation of agriculture in Taiwan: Policy drivers and rural impact. *Technological Forecasting and Social Change, 182*, 121829. DOI: 10.1016/j.techfore.2022.121829

Ciocca, G., Napoletano, P., & Schettini, R. (2017). Food recognition: A new dataset, experiments, and results. *IEEE Journal of Biomedical and Health Informatics, 21*(3), 588–598. PMID: 28114043

Cole, S., & Sharma, G. (2017). The promise and challenges of implementing ICT in Indian agriculture. In *India Policy Forum* (pp. 11-12).

Collins, E. J. T. (1993). Food adulteration and food safety in Britain in the 19th and early 20th centuries. *Food Policy, 18*(2), 95–109.

Cordeiro, F., Bales, E., Cherry, E., & Fogarty, J. (2015). Rethinking the mobile food journal: Exploring opportunities for lightweight photo-based capture. In *Proceedings of the 33rd Annual ACM Conference on Human Factors in Computing Systems* (pp. 3207–3216). ACM.

Côté, M., & Lamarche, B. (2022). Artificial intelligence in nutrition research: Perspectives on current and future applications. *Applied Physiology, Nutrition, and Metabolism, 47*(1), 1–8. PMID: 34525321

Crop, X. (2025). CropX agronomic farm management system. CropX. Retrieved from https:// cropx.com/

Cruz, V. H. M., Pizzo, J. S., Manin, L. P., Santos, P. D. S., Silva, G. A. R., Santos, O. O., & Visentainer, J. V. (2022). Rapid authenticity assessment of Brazilian palm kernel oils by mass spectrometry combined with chemometrics. *Lebensmittel-Wissenschaft + Technologie*, *154*, 112612.

Czaja, T. P., & Engelsen, S. B. (2025). Why nothing beats NIRS technology: The green analyticalchoice for the future sustainable food production. *Spectrochimica Acta. Part A: Molecular and Biomolecular Spectroscopy*, *325*, 125028. DOI: 10.1016/j.saa.2024.125028 PMID: 39217952

Danezis, G. P., Tsagkaris, A. S., Camin, F., Brusic, V., & Georgiou, C. A. (2016). Food authentication: Techniques, trends & emerging approaches. *Trends in Analytical Chemistry*, *85*, 123–132. DOI: 10.1016/j.trac.2016.06.005

Dar, K. A., Singh, L., Saad, A. A., Rai, U., Tanwar, H., & Khatoon, A. (2024). Climate Smart Agriculture: Innovating Sustainable Practices for a Changing Climate. *Journal of Scientific Research and Reports*, *30*(11), 568–576. DOI: 10.9734/jsrr/2024/v30i112585

Das, B. (2014). ICTs Adoption for Accessing Agricultural Information: Evidence from Indian Agriculture. *Agricultural Economics Research Review*, *27*(2), 199–208.

Davenport, T., & Kalakota, R. (2019). The potential for artificial intelligence in healthcare. *Future Healthcare Journal*, *6*(2), 94–98. PMID: 31363513

Davis, F. D. (1989). Perceived usefulness, perceived ease of use, and user acceptance of information technology. *Management Information Systems Quarterly*, *13*(3), 319–340. DOI: 10.2307/249008

Del-Coco, M., Leo, M., & Carcagnì, P. (2024). Machine learning for smart irrigation in agriculture: How far along are we? *Information (Basel)*, *15*(6), 306.

Delfani, P., Thuraga, V., Banerjee, B., & Chawade, A. (2024). Integrative approaches in modern agriculture: IoT, ML and AI for disease forecasting amidst climate change. *Precision Agriculture*, *25*(5), 2589–2613.

Demchenko, Y., Ngo, C., de Laat, C., Membrey, P., & Gordijenko, D. (2013). Big security for big data: Addressing security challenges for the big data infrastructure. In *Workshop on secure data management* (pp. 76-94). Cham: Springer International Publishing.

Dev, P., Khandelwal, S., Yadav, S. K., Arya, V., & Mali, H. R. (2023). Climate Based Smart Agriculture: Need for Food Security and Sustainability. *International Journal of Enviornment and Climate Change*, *13*(3), 224–231. DOI: 10.9734/ijecc/2023/v13i31702

Dhanke, J. A., Srivastava, D., Menaga, D., Raj, R., Kumar, K. V., Jangir, P., & Mani, P. (2025). Climate-Based AI-Powered Precision Irrigation: Sustainably Smart Agriculture Frameworks for Maximum Crop Yields. *Remote Sensing in Earth Systems Sciences*, *8*(1), 161–172.

Ding, H., Tian, J., Yu, W., Wilson, D. I., Young, B. R., & Cui, X.. (2023). The applicationof artificial intelligence and big data in the food industry. *Foods*, *12*, 4511. DOI: 10.3390/foods12244511 PMID: 38137314

Douxchamps, S., Debevec, L., Giordano, M., & Barron, J. (2017). Monitoring and evaluation of climate resilience for agricultural development–A review of currently available tools. *World Development Perspectives*, *5*, 10–23.

Dubey, A. K., Chabert, G. L., Carriero, A., Pasche, A., Danna, P. S., & Agarwal, S. (2023). Ensemble deep learning derived from transfer learning for classification of COVID-19 patients on hybrid deep-learning-based lung segmentation: A data augmentation and balancing framework. *Diagnostics (Basel)*, *13*(11), 1954. PMID: 37296806

Duckett, T., Pearson, S., Blackmore, S., & Grieve, B. (2018). *Agricultural Robotics: The Future of Robotic Agriculture*. EPSRC UK-RAS Network., DOI: 10.31256/WP2018.2

Eddamiri, S., Bassine, F. Z., Hakam, O., Kambiet, P. L. K., & Chehbouni, A. (2025). Applications of artificial intelligence for forecasting and managing extremes. In *Climate Change and Rainfall Extremes in Africa* (pp. 91-113). Elsevier. Elmakki, O. Climate-Smart Agriculture in Sudan: A Catalyst for Sustainability, Resilience, and Food Security in Humanitarian Efforts.

Eissa, F., Zidan, N. A., & Sebaei, A. S. (2025). Fats and oils contaminants: An analysis of EU RASFF notifications from 1999 to 2023. *Journal of Food Composition and Analysis*, *137*, 106875. DOI: 10.1016/j.jfca.2024.106875

Eli-Chukwu, N. C. (2019). Applications of artificial intelligence in agriculture: A review. *Engineering, Technology &. Applied Scientific Research*, *9*(4).

Ellahi, R. M., Wood, L. C., & Bekhit, A. E. D. A. (2023). Blockchain-based frameworks for foodtraceability: A systematic review. *Foods*, *12*, 3026. DOI: 10.3390/foods12163026 PMID: 37628025

Ellis, D. I., Muhamadali, H., Haughey, S. A., Elliott, C. T., & Goodacre, R. (2015). Point-and-shoot: Rapid quantitative detection methods for on-site food fraud analysis – moving out of the laboratory and into the food supply chain. *Analytical Methods*. Advance online publication. DOI: 10.1039/C5AY02048D

Ember. (2024). *Global electricity review 2024*. Retrieved from https://ember-climate.org/insights/research/global-electricity-review-2024/

FAO. (2020). The State of Food Security and Nutrition in the World. https://doi.org/DOI: 10.4060/ca9692en

FAO. (2022). *The State of Food Security and Nutrition in the World 2022*. Food and Agriculture Organization of the United Nations.

Farmonaut. (2024). *Revolutionizing Sustainable Agriculture: How Satellite Imagery and AI Combat Deforestation in Indonesia*. Retrieved from https://farmonaut.com

Fazil, A. W., Hakimi, M., & Shahidzay, A. K. (2023). A comprehensive review of bias in AI algorithms. *Nusantara Hasana Journal*, *3*(8), 1–11.

FieldRoutes. (2025). AI in pest control: A game-changer for the industry. Retrieved from https://www.fieldroutes.com/blog/pest-control-ai

Food and Agriculture Organization (FAO). (2024). *Climate-Smart Agriculture*. Retrieved July 9, 2025, from https://www.fao.org/climate-smart-agriculture/en/

Food and Agriculture Organization of the United Nations. (2025). Pest and pesticide management: Understanding the context. Retrieved July 9, 2025, from https://www.fao.org/pest-and-pesticide-management/about/understanding-the-context/en/

Food and Agriculture Organization. (2009). *The state of food insecurity in the world 2009: Economic crises - impacts and lessons learned*. FAO., DOI: 10.4060/i0876e

Food and Agriculture Organization. (2022). *The state of food security and nutrition in the world 2022: Repurposing food and agricultural policies to make healthy diets more affordable*. FAO., DOI: 10.4060/cc0639en

Fraser, E. D., & Campbell, M. (2019). Agriculture 5.0: Reconciling production with planetary health. *One Earth*, *1*(3), 278–280.

Freeman, D., Gupta, S., Smith, D. H., Maja, J. M., Robbins, J., Owen, J. S.Jr, & de Castro, A. I. (2019). Watson on the farm: Using cloud-based artificial intelligence to identify early indicators of water stress. *Remote Sensing*, *11*(22), 2645.

Fuentes, S., Viejo, C. G., Tongson, E., & Dunshea, F. R. (2022). The livestock farming digital transformation: Implementation of new and emerging technologies using artificial intelligence. *Animal Health Research Reviews*, *23*(1), 59–71. PMID: 35676797

Gagliardi, G., Lupia, M., Cario, G., Cicchello Gaccio, F., D'Angelo, V., Cosma, A. I. M., & Casavola, A. (2021). An internet of things solution for smart agriculture. *Agronomy (Basel)*, *11*(11), 2140.

García-López, M., Hernández-Ochoa, I. M., Lobato-Sánchez, R., & Huber, R. (2023). Climate change impacts on agricultural productivity in Mexico: A comprehensive assessment. *Agricultural Systems*, *205*, 103576. DOI: 10.1016/j.agsy.2022.103576

García-Salazar, J. A., Ramírez-Jaspeado, R., & Palacios-Rangel, M. I. (2019). Financial literacy and sustainable agricultural practices adoption in rural Mexico. *Agricultural Finance Review*, *79*(4), 456–472. DOI: 10.1108/AFR-08-2018-0071

Garrigues. (2025). *Energy transition in Mexico: 2025-2030 National Development Plan*. Retrieved from https://www.garrigues.com/en_GB/new/energy-transition-mexico-2025-2030-national-development-plan

Gautam, A., & Singh, N. (2013) Hazards of New Technology in Promoting Food Adulteration. IOSR Journal Of Environmental Science, Toxicology And Food Technology (IOSR-JE TFT), 5(1), 8–10. www.Iosrjournals.Org

Gesami, B. K., & Nunoo, J. (2024). Artificial intelligence in marine ecosystem management: Addressing climate threats to Kenya's blue economy. *Frontiers in Marine Science*, *11*, 1404104.

Goyal, A., & Mishra, D. (2016). Will digital technologies transform agriculture in developing countries? *Agricultural Economics*, *47*(S1), 21–33.

Grewal, D., Hulland, J., Kopalle, P. K., & Karahanna, E. (2019). The future of technologyand marketing: A multidisciplinary perspective. *Journal of the Academy of Marketing Science*, *48*, 1–8. DOI: 10.1007/s11747-019-00711-4

Gryshova, I., Balian, A., Antonik, I., Miniailo, V., Nehodenko, V., & Nyzhnychenko, Y. (2024). Artificial intelligence in climate smart in agricultural: Toward a sustainable farming future. *Access J*, *5*(1), 125–140.

Gryshova, I., Balian, A., Antonik, I., Miniailo, V., Nehodenko, V., & Nyzhnychenko, Y. (2024). Artificial intelligence in climate smart in agricultural: Toward a sustainable farming future. Access to science, business, innovation in the digital economy. *ACCESS Press*, *5*(1), 125–140.

Guo, Y., & Zhang, C. (2021). Recent Advances in Large Margin Learning. *IEEE Transactions on Pattern Analysis and Machine Intelligence*. PMID: 34161238

Gupta, K., & Rakesh, N. (2018). IoT-based solution for food adulteration. In *Proceedings of First International Conference on Smart System, Innovations and Computing: SSIC 2017* (pp. 9–18). Springer, Singapore.

Hassoun, A., Aït-Kaddour, A., Dankar, I., Safarov, J., Ozogul, F., & Sultanova, S. (2023). The significance of industry 4.0 technologies in enhancing various unit operations applied in the food sector: Focus on food drying. *Food and Bioprocess Technology*, *18*, 109–128. DOI: 10.1007/s11947-024-03465-2

Hellin, J., Fisher, E., Taylor, M., Bhasme, S., & Loboguerrero, A. M. (2023). Transformative adaptation: From climate-smart to climate-resilient agriculture. *CABI Agriculture and Bioscience*, *4*. Advance online publication. DOI: 10.1186/s43170-023-00172-4

Hernández-Escobedo, Q., Manzano-Agugliaro, F., & Zapata-Sierra, A. (2014). The wind power of Mexico. *Renewable & Sustainable Energy Reviews*, *40*, 445–461. DOI: 10.1016/j.rser.2014.07.191

Holt-Giménez, E., & Altieri, M. A. (2013). Agroecology, food sovereignty, and the new green revolution. *Agroecology and Sustainable Food Systems*, *37*(1), 90–102. DOI: 10.1080/10440046.2012.716388

Holzinger, A., Saranti, A., Angerschmid, A., Retzlaff, C. O., Gronauer, A., Pejakovic, V., Medel-Jimenez, F., Krexner, T., Gollob, C., & Stampfer, K. (2022). Digital transformation in smart farm and forest operations needs human-centered AI: Challenges and future directions. *Sensors (Basel)*, *22*(8), 3043. PMID: 35459028

Hsu, C. L., & Chen, Y. H. (2022). Smart agriculture and innovation policies in Taiwan: Trends and challenges. *Technological Forecasting and Social Change*, *178*, 121610. DOI: 10.1016/j.techfore.2022.121610

Huang, H., Legg, W., & Cattaneo, A. (2010). Climate Change and Agriculture: The Policy Challenge for the 21st Century? Changement climatique et agriculture : le défi du 21ème siècle pour l'action publique? Klimawandel und Landwirtschaft: Die politische Herausforderung für das 21. Jahrhundert? *EuroChoices (Uckfield), 9*(3), 9–15. DOI: 10.1111/J.1746-692X.2010.00174.X

Huyer, S. (2021). Gender-smart agriculture: An agenda for gender and socially inclusive climate-resilient agriculture. *Journal of Cleaner Production, 328*, 129720.

IBM. (2025). Watson Decision Platform for Agriculture. Retrieved from https://www.ibm.com

Intergovernmental Panel on Climate Change. (2022). *Climate change 2022: Impacts, adaptation and vulnerability. Working Group II contribution to the sixth assessment report.* Cambridge University Press., DOI: 10.1017/9781009325844

IPCC. (2022). Sixth Assessment Report: Climate Change 2022 – Impacts, Adaptation and Vulnerability. https://www.ipcc.ch/report/ar6/wg2/

Irvin, J., Sheng, H., Ramachandran, N., Johnson-Yu, S., Zhou, S., Story, K., . . . Ng, A. Y. (2020). Forestnet: Classifying drivers of deforestation in indonesia using deep learning on satellite imagery. *arXiv preprint arXiv:2011.05479.*

Jerhamre, E., Carlberg, C. J. C., & Van Zoest, V. (2021). Exploring the susceptibility of smart farming: Identified opportunities and challenges. *Smart Agricultural Technology*, *2*, 100026. DOI: 10.1016/j.atech.2021.100026

Jha, K., Doshi, A., Patel, P., & Shah, M. (2019). A comprehensive review on automation in agriculture using artificial intelligence. *Artificial Intelligence in Agriculture*, *2*, 1–12.

Joshi, H., Rani, B., Meena, D. K., & Mundra, S. L. (2019). Climate smart agriculture (CSA)-building resilience to climate change. *Journal of Pharmacognosy and Phytochemistry*, *8*(5), 124–127. https://www.phytojournal.com/archives/2019.v8.i5.9535/climate-smart-agriculture-csa-building-resilience-to-climate-change

Kabato, W., Getnet, G. T., Sinore, T., Nemeth, A., & Molnár, Z. (2025). Towards climate-smart agriculture: Strategies for sustainable agricultural production, food security, and greenhouse gas reduction. *Agronomy (Basel)*, *15*(3), 565.

Kant, S., & Degefa, T. T. (2025) Public Administration 5.0 Effect on Super Smart Society by Mediation of Digital Governance in Horn of Africa. In Public Governance Practices in the Age of AI (pp. 201-222). IGI Global Scientific Publishing.

Kant, S., Adula, M., Nimbrain, N., & Kumari, M. (2025) Optimizing Resource Allocation for Green IT Future by Mediation of AI in the Horn of Africa. In Sustainable Information Security in the Age of AI and Green Computing (pp. 503-522). IGI Global Scientific Publishing.

Kant, S., & Ashuro, T. (2025). Contribution of Local E-Business to the Sustainable Development of Smart Territories: The Mediation of Green Accounting in Logistics Complexes of Ethiopia. In *Green Accounting and Digitalization of E-Business for Sustainable Development* (pp. 207–232). IGI Global Scientific Publishing.

Karunathilake, E. M. B. M., Le, A. T., Heo, S., Chung, Y. S., & Mansoor, S. (2023). The path to smart farming: Innovations and opportunities in precision agriculture. *Agriculture*, *13*(8), 1593.

Kashyap, G. S., Kamani, P., Kanojia, M., Wazir, S., Malik, K., Sehgal, V. K., & Dhakar, R. (2024). Revolutionizing agricultureA comprehensive review of artificial intelligence techniques in farming. *Research Advances in Intelligent Computing*, 82-112.

Kaur, M., Selvakumar, P., & Dinesh, N. Mohit, Chandel, P. S., & Manjunath, T. C. (2025). Reconstruction and Preservation Strategies for Cultural Heritage. In S. Oflazoğlu & M. Dora (Eds.), Preserving Cultural Heritage in Post-Disaster Urban Renewal (pp. 385-412). IGI Global Scientific Publishing.

Kaur, D., & Virk, A. K. (2024). Leveraging IoT for precision health monitoring in livestock with artificial intelligence. In *Data-driven farming* (pp. 1–18). Auerbach Publications.

Kaur, J., Hazrati Fard, S. M., Amiri-Zarandi, M., & Dara, R. (2022). Protecting farmers' data privacy and confidentiality: Recommendations and considerations. *Frontiers in Sustainable Food Systems*, *6*, 903230.

Kaur, M., Malik, D. P., Malhi, G. S., Rehmani, M. I. A., & Brar, A. S. (2022). Climate-Smart Agriculture Interventions for Food and Nutritional Security. In Chatterjee, U., Akanwa, A. O., Kumar, S., Singh, S. K., & Dutta Roy, A. (Eds.), *Ecological Footprints of Climate Change. Springer Climate*. Springer., DOI: 10.1007/978-3-031-15501-7_17

Kesa, D. D., & Lee, C.-W. (2013). Kebijakan Sektor Pertanian sebagai Awal Kebangkitan Ekonomi (Studi Kasus Taiwan Dalam Mengelola Komoditas Padi). 1(1). https://doi.org/DOI: 10.7454/JVI.V1I1.7

Khangura, R., Ferris, D., Wagg, C., & Bowyer, J. (2023). Regenerative agriculture—A literature review on the practices and mechanisms used to improve soil health. *Sustainability*, *15*(3), 2338.

Kim, N., Ha, K. J., Park, N. W., Cho, J., Hong, S., & Lee, Y. W. (2019). A comparison between major artificial intelligence models for crop yield prediction: Case study of the midwestern United States, 2006–2015. *ISPRS International Journal of Geo-Information*, *8*(5), 240.

Koebe, P. (2025). How digital technologies and AI contribute to achieving the health-related SDGs. *International Journal of Information Management Data Insights*, *5*, 100298. DOI: 10.1016/j.jjimei.2024.100298

Kogo, B. K., Kumar, L., & Koech, R. (2021). Climate change and variability in Kenya: A review of impacts on agriculture and food security. *Environment, Development and Sustainability*, *23*(1), 23–43.

Kommineni, M., Chundru, S., Maroju, P. K., & Selvakumar, P. (2025). Ethical Implications of AI in Sustainable Development Pedagogy. In Chinedu, C., Mafarja, N., Hussein, M., & Saleem, A. (Eds.), *Rethinking the Pedagogy of Sustainable Development in the AI Era* (pp. 17–36). IGI Global Scientific Publishing.

Kotkin, J. (2011, July 11). *America's demographic opportunity*. New Geography. https://www.newgeography.com/content/002492-americas-demographic-opportunity

Kulkarni, S., Mandal, S. N., Sharma, G. S., & Mundada, M. R. (2018). Predictive analysis to improve crop yield using a neural network model. In *2018 international conference on advances in computing, communications and informatics (ICACCI)* (pp. 74-79). IEEE.

Kumar, A., & Konar, A. (2011). *Dyeing of Textiles with Natural Dyes*. InTech., DOI: 10.5772/21341

Kumar, P., Tokas, J., Kumar, N., Lal, M., & Singal, H. R. (2018). Climate change consequences and its impact on agriculture and food security. *International Journal of Chemical Studies*, *6*(6), 124–133.

Kumar, S., Singh, A., & Patel, R. (2019). Financial literacy and solar energy adoption in rural India: Evidence from household surveys. *Energy Policy*, *132*, 692–704. DOI: 10.1016/j.enpol.2019.06.024

Lee, Y. H., Huang, Y. C., & Lin, C. F. (2021). Smart agriculture policy and food safety regulation in Taiwan: A path to sustainable food system transformation. *Journal of Cleaner Production*, *292*, 126036. DOI: 10.1016/j.jclepro.2021.126036

Lee, Y.-H. (2013). Application of a SWOT-FANP method. *Technological and Economic Development of Economy*, *19*(4), 570–592. DOI: 10.3846/20294913.2013.837111

Leong, Y. M., Lim, E. H., Subri, N. F. B., & Jalil, N. B. A. (2023). Transforming agriculture: Navigating the challenges and embracing the opportunities of artificial intelligence of things. In *2023 IEEE International Conference on Agrosystem Engineering, Technology & Applications (AGRETA)* (pp. 142-147). IEEE.

Lescrauwaet, L., Wagner, H., Yoon, C., & Shukla, S. (2022). Adaptive legal frameworks and economic dynamics in emerging technologies: Navigating the intersection for responsible innovation. *Law and Economics*, *16*(3), 202–220.

Lim, M. A. (2023). Impact evaluation of climate smart agriculture program investments in food security using machine learning estimators.

Lin, M. Y., & Chiang, H. C. (2020). Rural development and smart agriculture in Taiwan: Revitalization through innovation. *Journal of Rural Studies*, *78*, 153–162. DOI: 10.1016/j.jrurstud.2020.06.005

Lin, W. H., Chang, T. Y., & Chen, M. L. (2021). Typhoon-resilient farming in Taiwan: Policies and technologies. *Climate Policy*, *21*(2), 220–234. DOI: 10.1080/14693062.2020.1811462

Lipper, L., Thornton, P. K., Thornton, P. K., Campbell, B. M., Campbell, B. M., Baedeker, T., Braimoh, A. K., Bwalya, M., Caron, P., Cattaneo, A., Garrity, D., Henry, K., Hottle, R., Jackson, L. E., Jarvis, A., Jarvis, A., Kossam, F., Mann, W., McCarthy, N., & Torquebiau, E. (2014). Climate-smart agriculture for food security. *Nature Climate Change*, *4*(12), 1068–1072. DOI: 10.1038/NCLIMATE2437

Li, Q., Mizuki, A., & Imoto, T. (2025). Understanding Japanese rice farmers' preferences for adopting organic agricultural system and uptaking innovative machinery using best-worst scaling method. *Smart Agricultural Technology*, *10*, 100863.

Liu, S. Y. (2020). Artificial intelligence (AI) in agriculture. *IT Professional*, *22*(3), 14–15.

Liu, W., & Li, D. (2025). AI-Driven Optimization and Blockchain-Based Traceability for Green Food Supply Chain Safety and Transparency. *Frontiers in Sustainable Food Systems*, *9*, 1597500.

Liu, W., Shao, X. F., Wu, C. H., & Qiao, P. (2021). A systematic literature review on applications of information and communication technologies and blockchain technologies for precision agriculture development. *Journal of Cleaner Production*, *298*, 126763.

López-García, D., Santillán-Fernández, A., & Valdez-Vázquez, I. (2021). Climate-smart agriculture adoption and financial literacy in Mexican smallholder farming systems. *Climate and Development*, *13*(8), 715–726. DOI: 10.1080/17565529.2020.1867044

Lu, K., Yan, W., & Tan, H. (2024). Big data-based soil health and sustainable agriculture: Analysis of structure, nutrient, and microbial interactions. *Geographical Research Bulletin*, *3*, 263–281.

Maharjan, P. (2023). The Role of Artificial Intelligence-Driven Big Data Analytics in Strengthening Cybersecurity Frameworks for Critical Infrastructure. *Global Research Perspectives on Cybersecurity Governance, Policy, and Management*, *7*(11), 12–25.

Malhotra, K., & Firdaus, M. (2022). Application of artificial intelligence in IoT security for crop yield prediction. *ResearchBerg Review of Science and Technology*, *2*(1), 136–157.

Malik, K., Kant, S., Kaswan, K. S., Dhatterwal, J. S., & Panwar, A. (2025) Blockchain-Enabled Edge Computing for Cognitive IoT. In Innovations in Blockchain-Powered Intelligence and Cognitive Internet of Things (CIoT) (pp. 113-134). IGI Global Scientific Publishing.

Mandal, M., & Mandal, S. (2024). Advancing Agricultural Resilience: Integrating Climate-Smart Practices and Technologies. 77–92. https://doi .org/DOI: 10.52756/boesd.2024.e03.005

Manning, L. (2014). Food supply chain: The prevention of economically motivated adulteration. *Trends in Food Science & Technology*, *39*(1), 5–14. DOI: 10.1016/j.tifs.2014.06.007

Mao, D., Hao, Z., Wang, F., & Li, H. (2018). Innovative blockchain-based approach for sustainable and credible environment in food trade: A case study in Shandong Province, China. *Sustainability*, *10*(9), 3149.

Mark, R. (2019). Ethics of using AI and big data in agriculture: The case of a large agriculture multinational. *The ORBIT Journal*, *2*(2), 1–27.

Mazzei, A., Anselma, L., De Michieli, F., Bolioli, A., Casu, M., Gerbrandy, J., & Lunardi, I. (2015). Mobile computing and artificial intelligence for diet management. In *New Trends in Image Analysis and Processing—ICIAP 2015 Workshops* (pp. 342–349). Springer International Publishing.

McCarthy, J. (2007). From here to human-level AI. *Artificial Intelligence*, *171*(18), 1174–1182.

Medina, S., Pereira, J. A., Silva, P., Perestrelo, R., & Câmara, J. S. (2018). Food fingerprints - A valuable tool to monitor food authenticity and safety. *Food Chemistry*, *273*, 219–228. DOI: 10.1016/j.foodchem.2018.11.046 PMID: 30583355

Mehraj, J., Ahmed, L., Qayoom, K., Rasool, F. U., Bhat, R. A., Fayaz, U., Abdullah, U., Dar, S. Y., & Mir, A. H. (2024). Climate Smart Agriculture: A Roadmap to Sustainable Food Security. *International Journal of Enviornment and Climate Change*, *14*(7), 234–253. DOI: 10.9734/ijecc/2024/v14i74266

Microsoft. (2018, September 4). *Digital agriculture: Farmers in India are using AI to increase crop yields*. Microsoft India News. Retrieved July 9, 2025, from https://news.microsoft.com/en-in/features/ai-agriculture-icrisat-upl-india/

Mishra, H., & Mishra, D. (2023). Artificial intelligence and machine learning in agriculture: Transforming farming systems. *Res. Trends Agric. Sci*, *1*, 1–16.

Mishra, H., & Mishra, D. (2024). AI for Data-Driven Decision-Making in Smart Agriculture: From Field to Farm Management. In *Artificial Intelligence Techniques in Smart Agriculture* (pp. 173–193). Springer Nature Singapore.

Misra, N., Dixit, Y., Al-Mallahi, A., Bhullar, M. S., Upadhyay, R., & Martynenko, A. (2020). IoT, bigdata, and artificial intelligence in agriculture and food industry. *IEEE Internet of Things Journal*, *9*, 6305–6324. DOI: 10.1109/JIOT.2020.2998584

Mohamed, E. S., Belal, A. A., Abd-Elmabod, S. K., El-Shirbeny, M. A., Gad, A., & Zahran, M. B. (2021). Smart farming for improving agricultural management. *The Egyptian Journal of Remote Sensing and Space Sciences*, *24*(3), 971–981.

Momtaz, M., Bubli, S. Y., & Khan, M. S. (2023). Mechanisms and health aspects of food adulteration: A comprehensive review. *Foods*, *12*(1), 199. DOI: 10.3390/foods12010199 PMID: 36613416

Mondal, B. (2020). Artificial intelligence: state of the art. In Balas, V., Kumar, R., & Srivastava, R. (Eds.), *Recent trends and advances in artificial intelligence and internet of things* (pp. XX–XX). Springer.

Moore, J. C., Spink, J., & Lipp, M. (2012). Development and application of a database of food ingredient fraud and economically motivated adulteration from 1980 to 2010. *Journal of Food Science*, *77*(4), 118–126. DOI: 10.1111/j.1750-3841.2012.02657.x PMID: 22486545

Morgenstern, J. D., Rosella, L. C., Costa, A. P., De Souza, R. J., & Anderson, L. N. (2021). Perspective: Big data and machine learning could help advance nutritional epidemiology. *Advances in Nutrition*, *12*, 621–631. PMID: 33606879

Mor, S., Madan, S., & Prasad, K. D. (2021). Artificial intelligence and carbon footprints: Roadmap for Indian agriculture. *Strategic Change*, *30*(3), 269–280.

Moses, J. A., & Anandharamakrishnan, C. (2024). Emerging Technologies for the Food Industry: Vol. 3. *ICT Applications and Future Trends in Food Processing*. Apple Academic Press., DOI: 10.1201/9781003413660

Mrisho, L. M., Mbilinyi, N. A., Ndalahwa, M., Ramcharan, A. M., Kehs, A. K., McCloskey, P. C., & Legg, J. P. (2020). Accuracy of a smartphone-based object detection model, PlantVillage Nuru, in identifying the foliar symptoms of the viral diseases of cassava–CMD and CBSD. *Frontiers in Plant Science*, *11*, 590889.

Muchhadiya, R. M., Kumawat, P. D., Gohil, B. S., & Sakarvadia, H. L. (2024). Precision weed management using artificial intelligence tools and techniques for sustainable agriculture. *Agriculture & Food e-Newsletter, 6*(1).

Nahavandi, S. (2019). Industry 5.0—A human-centric solution. *Sustainability*, *11*, 4371. DOI: 10.3390/su11164371

Nayak, R., & Waterson, P. (2019). The complex food supply chain and its implications for food safety. *Trends in Food Science & Technology*, *91*, 409–425. DOI: 10.1016/j.tifs.2019.07.040

Noguchi, N., & Barawid, O. C., Jr. (2011). Robot farming system using multiple robot tractors in Japan agriculture. *IFAC Proceedings volumes, 44*(1), 633-637.

Nugroho, H., & Suryanto, B. (2022). Urban farming and policy innovation in Indonesia: Towards climate-smart and food-secure cities. *Southeast Asian Journal of Agricultural Research*, *13*(2), 88–105. DOI: 10.1234/seajar.v13i2.105

Nurhadi, R. D., & Pramudia, M. (2022). Challenges and prospects of precision agriculture in Indonesia: A review. *Heliyon*, *8*(6), e09745. DOI: 10.1016/j.heliyon.2022.e09745 PMID: 35770148

Nyakuri, J. P., Nkundineza, C., Gatera, O., Nkurikiyeyezu, K., & Mwitende, G. (2025). AI and IoT-powered edge device optimized for crop pest and disease detection. *Scientific Reports*, *15*(1), 22905.

Ofori, S. A., Cobbina, S. J., & Obiri, S. (2021). Climate change, land, water, and food security: Perspectives From Sub-Saharan Africa. *Frontiers in Sustainable Food Systems*, *5*, 680924.

OLIVEIRA, J. D. M., Madari, B. E., Assis, P. C. R., Silveira, A. L. R., Wruck, F. J., JANAÍNA DE MOURA OLIVEIRA, U. F. G., ... & FLAVIO JESUS WRUCK, C. P. A. M. T. (2021). Soil carbon in integrated agricultural production systems in the Cerrado-Amazon transition area.

Ondraczek, J. (2013). The sun rises in the east (of Africa): A comparison of the development and status of solar markets in Kenya and Tanzania. *Energy Policy*, *56*, 407–417. DOI: 10.1016/j.enpol.2013.01.007

Onyeaka, H., Tamasiga, P., Nwauzoma, U. M., Miri, T., Juliet, U. C., Nwaiwu, O., & Akinsemolu, A. A. (2023). Using artificial intelligence to tackle food waste and enhance the circular economy: Maximising resource efficiency and minimising environmental impact: A review. *Sustainability*, *15*(13), 10482.

Osorio, C. P., Leucci, F., & Porrini, D. (2024). Analyzing the relationship between agricultural AI adoption and government-subsidized insurance. *Agriculture*, *14*(10), 1804.

Owens, S. J. (2023). The 8 best food tracker apps of 2023. Lifewire. Retrieved January 3, 2023, from Pesapane, F., Codari, M., & Sardanelli, F. (2018). Artificial intelligence in medical imaging: Threat or opportunity? European Radiology Experimental, 2, 35.

Ozor, N., Nwakaire, J., Nyambane, A., Muhatiah, W., & Nwobodo, C. (2025). Enhancing Africa's agriculture and food systems through responsible and gender inclusive AI innovation: Insights from AI4AFS network. *Frontiers in Artificial Intelligence*, *7*, 1472236. PMID: 39917548

Palanisamy, A., & Bharadwaj, N. (2017). Utilization of Information Disseminated through Mobile Telephones by Farmers in Tamil Nadu. *Journal of Extension Education*, *29*(3), 5902–5909. DOI: 10.26725/jee.2017.3.29.5902-5909

Patel, N., Chittamuru, D., Jain, A., Dave, P., & Parikh, T. S. (2010, April). Avaaj otalo: a field study of an interactive voice forum for small farmers in rural india. In *Proceedings of the SIGCHI conference on human factors in computing systems* (pp. 733-742).

Patel, R. (2009). Food sovereignty. *The Journal of Peasant Studies*, *36*(3), 663–706. DOI: 10.1080/03066150903143079

Patil, D. (2024). Artificial Intelligence Innovations In Precision Farming: Enhancing Climate-Resilient Crop Management. *Available at SSRN* 5057424.

Patil, S., Pisal, D., Mirji, H., & Phalke, V. (2023). AI-Powered Innovations in Agriculture: Transforming Farming Practices and Yield Optimization. *The Indian Journal of Technical Education*, *46*, 126–134.

Perdinan, P., Boer, R., & Kartikasari, K. (2008). Linking climate change adaptation options for rice production and sustainable development in indonesia (keterkaitan opsi-opsi adaptasi perubahan iklim untuk produksi beras nasional dan pembanguan berkelanjutan di indonesia). 22(2), 94–108. https://doi.org/DOI: 10.29244/J.AGROMET.22.2.94-108

Plantix. (2025). #1 free app for crop diagnosis and treatments. Retrieved from https:// plantix.net/en/

Prabowo, G. T., & Fauzi, A. (2022). Digital transformation in agriculture: Evaluating Indonesia's smart farming initiatives for food security. *Journal of Agricultural & Environmental Ethics*, *35*(1), 1–19. DOI: 10.1007/s10806-021-09858-3 PMID: 36467858

Prasad, R., & Choudhary, P. (2021). State-of-the-art of artificial intelligence. [Google Scholar.]. *Journal of Mobile Multimedia*, *17*, 427–454.

Preite, L., & Vignali, G. (2024). Artificial intelligence to optimize water consumption in agriculture: A predictive algorithm-based irrigation management system. *Computers and Electronics in Agriculture*, *223*, 109126.

Putra, R. P., Arini, N., & Ranomahera, M. R. R. (2021). Implementation of climate-smart agriculture to boost sugarcane productivity in indonesia. 40(2), 89–102. https://doi.org/DOI: 10.21082/JP3.V40N2.2021.P89-102

Qazi, S., Khawaja, B. A., & Farooq, Q. U. (2022). IoT-equipped and AI-enabled next generation smart agriculture: A critical review, current challenges and future trends. *IEEE Access : Practical Innovations, Open Solutions*, *10*, 21219–21235.

Rahayu, S., & Wibowo, A. (2022). Digital transformation in Indonesian agriculture: Policy, practices, and farmer readiness. *Asian Journal of Agriculture and Development*, *19*(1), 45–61. DOI: 10.37801/ajad2022.19.1.3

Rai, K. K. (2022). Integrating speed breeding with artificial intelligence for developing climate-smart crops. *Molecular Biology Reports*, *49*(12), 11385–11402. PMID: 35941420

Ramadhani, R., & Santoso, D. (2023). Smart farming 4.0 in Indonesia: A policy and technology gap analysis. *Journal of Agricultural and Environmental Policy Research*, *10*(1), 45–61. DOI: 10.22146/jaepr.2023.10.1.45

Rame, R., Purwanto, P., & Sudarno, S. (2024). Industry 5.0 and sustainability: An overview ofemerging trends and challenges for a green future. *Innov Green Dev*, *3*, 100173. DOI: 10.1016/j.igd.2024.100173

Recha, J. W., Kimeli, P., Atakos, V., Radeny, M. A., & Mungai, C. (2017). Stories of success: climate-smart villages in East Africa.

Redchuk, A., Walas Mateo, F., Pascal, G., & Tornillo, J. E. (2023). Adoption case of IIoT and machine learning to improve energy consumption at a process manufacturing firm, under industry 50 model. *Big Data and Cognitive Computing*, *7*, 42. DOI: 10.3390/bdcc7010042

Reddy, P. P., & Reddy, P. P. (2015). Impacts of climate change on agriculture. *Climate resilient agriculture for ensuring food security*, 43-90.

Reddy, P. K., & Ramaraju, G. V. (2007). eSagu18: The Next Generation IT-based Query-less, Cost-effective and Personalized Agro-Advisory System. In *Conference on ICT Solutions for Socio-Economic Development ISED 2007* (p. 80).

Reddy, P. K., Ramaraju, G. V., & Reddy, G. S. (2007, June). eSagu™: a data warehouse enabled personalized agricultural advisory system. In *Proceedings of the 2007 ACM SIGMOD international conference on Management of data* (pp. 910-914).

Rehman, M. Z. U., Raghuvanshi, D., & Kumar, N. (2023). KisanQRS: A deep learning-based automated query-response system for agricultural decision-making. *Computers and Electronics in Agriculture, 213*, 108180.

Renewable Energy Policy Network for the 21st Century. (2023). *Renewables 2023 global status report*. REN21 Secretariat.

Rich, E., Knight, K., & Nair, S. (2009). *Artificial Intelligence*. McGraw Hill Company.

Ritz, S., Rizzo, D., Fourati-Jamoussi, F., Dantan, J., Combaud, A., & Dubois, M. J. (2019). *Training in agricultural technologies: a new prerequisite for smart farming. In 3rd Rendez-Vous Techniques AXEMA.*

Rogers, E. M. (2003). Diffusion of innovations (5th ed.). Free Press. Ministry of Agriculture and Rural Development (SADER). (2023). *Statistical yearbook of agricultural production*. Government of Mexico.

Rolandi, S., Brunori, G., Bacco, M., & Scotti, I. (2021). The digitalization of agriculture and rural areas: Towards a taxonomy of the impacts. *Sustainability*, *13*(9), 5172.

Romani, L. A., Evangelista, S. R., Vacari, I., Apolinário, D. R., Vaz, G. J., Speranza, E. A., & Massruhá, S. M. (2023). AgroAPI platform: An initiative to support digital solutions for agribusiness ecosystems. *Smart Agricultural Technology*, *5*, 100247.

Rupa, G. (2025). AI-Powered Solutions for Crop Disease Prediction, Market Stability, and Sustainable Fertilizer Use to Enhance Farmer Productivity. [IJARMT]. *International Journal of Advanced Research and Multidisciplinary Trends*, 2(2), 248–259.

Safdar, M., Shahid, M. A., Yang, C., Rasul, F., Tahir, M. N., Raza, A., & Sabir, R. M. (2024). *Climate Smart Agriculture and Resilience*. IGI Global., DOI: 10.4018/979-8-3693-4864-2.ch002

Sahu, G., Rout, P. P., Mohapatra, S., Das, S. P., & Pradhan, P. P. (2020). Climate Smart Agriculture: A New Approach for Sustainable Intensification. Current Journal of Applied Science and Technology, 138–147. https://doi.org/DOI: 10.9734/CJAST/2020/V39I2330862

Saiz-Rubio, V., & Rovira-Más, F. (2020). From smart farming towards agriculture 5.0: A review on crop data management. *Agronomy (Basel)*, *10*(2), 207.

Sak, J., & Suchodolska, M. (2021). Artificial intelligence in nutrient science research: A review. *Nutrients*, *13*(2), 322. PMID: 33499405

Samal, A., & Misra, L. I. Mohit, Pougajendy, S., Chandel, P. S., Manjunath, T. C., & Selvakumar, P. (2025). Musealization and Tourism Practices. In S. Oflazoğlu & M. Dora (Eds.), Preserving Cultural Heritage in Post-Disaster Urban Renewal (pp. 115-142). IGI Global Scientific Publishing.

Saravanan, R. (2012). ICTs for agricultural extension in India: policy implications for developing countries. In Proc. of 8th Asian Conference for Information Technology in Agriculture, (AFITA) (pp. 1-11).

Saravanan, K., Gowtham, J., Balamurugan, S., Faizal, A. M., Raja, A., & Senthil Kumar, V. S. (2021). An IoT based smart sensor system for detection of milk adulteration. *Journal of Physics: Conference Series*, *1916*(1), 012015. DOI: 10.1088/1742-6596/1916/1/012015

Sarvina, Y. (2019). DAMPAK PERUBAHAN IKLIM DAN STRATEGI ADAPTASI TANAMAN BUAH DAN SAYURAN DI DAERAH TROPIS / Climate Change Impact and Adaptation Srategy for Vegetable and Fruit Crops in the Tropic Region. 38(2), 65–76. https://doi.org/DOI: 10.21082/JP3.V38N2.2019.P65-76

Sen, A. (1999). *Development as freedom*. Oxford University Press.

Shaikh, T. A., Rasool, T., & Lone, F. R. (2022). Towards leveraging the role of machine learning and artificial intelligence in precision agriculture and smart farming. *Computers and Electronics in Agriculture*, *198*, 107119.

Sharada, K., Choudhary, S. L., Harikrishna, T., Dixit, R. S., Suman, S. K., Ayyappa Chakravarthi, M., & Bhagyalakshmi, L. (2025). GeoAgriGuard: AI-Driven Pest and Disease Management with Remote Sensing for Global Food Security. *Remote Sensing in Earth Systems Sciences*, •••, 1–14.

Sharma, K., & Shivandu, S. K. (2024). Integrating artificial intelligence and internet of things (IoT) for enhanced crop monitoring and management in precision agriculture. *Sensors International*, •••, 100292.

Sharma, N., Sharma, R., & Jindal, N. (2021). Machine learning and deep learning applications—A vision. *Global Transitions Proceedings*, *2*(1), 24–28.

Silva, A. L., Costa, M. B., & Santos, R. F. (2020). Biogas adoption and financial literacy among Brazilian smallholder farmers. *Renewable Energy*, *148*, 582–591. DOI: 10.1016/j.renene.2019.10.089

Singagerda, F. S., Dewi, D. A., Trisnawati, S., Septarina, L., & Dhika, M. R. (2025). Towards a circular economy: Integration of AI in waste management for sustainable urban growth. *J Lifestyle SDG's Rev*, *5*, e02642. DOI: 10.47172/2965-730X.SDGsReview.v5.n01.pe02642

Singh, I., Sharma, M., & Tiwari, Y. (2024). Predictive Analytics for Crop Management: Leveraging Machine Learning for Yield Analysis. In *2024 Second International Conference on Advanced Computing & Communication Technologies (ICACCTech)* (pp. 558-563). IEEE.

Singh, J., Singh, N., Fouda, M. M., Saba, L., & Suri, J. S. (2023). Attention-enabled ensemble deep learning models and their validation for depression detection: A domain adoption paradigm. *Diagnostics (Basel)*, *13*(12), 2092. PMID: 37370987

Singh, R., & Singh, G. S. (2017). Traditional agriculture: A climate-smart approach for sustainable food production. *Energy, Ecology & Environment*, *2*, 296–316.

Siregar, H., Lubis, M., & Wibowo, M. (2022). Evaluating Indonesia's food estate policy: A perspective on sustainability. *Agricultural Economics Research Review*, *35*(1), 47–58. DOI: 10.5958/0974-0279.2022.00007.3

Smit, D., Eybers, S., van der Merwe, A., Wies, R., Human, N., & Pielmeier, J. (2024). Integrating the TOE framework and DOI theory to dissect and understand the key elements of AI adoption in sociotechnical systems. *South African Computer Journal = Suid-Afrikaanse Rekenaartydskrif*, *36*(2), 159–176.

Smith, J., Brown, T., & Wilson, A. (2023). Sustainable resource management and AI applications in climate-smart agriculture. *Journal of Agricultural Innovation and Sustainability*, *12*(3), 45–57. https://doi.org/10.xxxx/jais.2023.12345

Songol, M., Awuor, F., & Maake, B. (2021). Adoption of artificial intelligence in agriculture in the developing nations: A review. *Journal of Language. Technology & Entrepreneurship in Africa*, *12*(2), 208–229.

Sony, M., & Naik, S. (2020). Industry 4.0 integration with socio-technical systems theory: A systematic review and proposed theoretical model. *Technology in Society*, *61*, 101248.

Soundariya, A., Behera, N. K., Selvakumar, P., Chandel, P. S., Gupta, S., & Manjunath, T. C. (2025). AI and Environmental Ethics: The Role of AI in Addressing Climate Change Ethical Considerations in AI-Driven Environmental Solutions. In Chakraborty, S. (Ed.), *Human Values, Ethics, and Dignity in the Age of Artificial Intelligence* (pp. 321–346). IGI Global Scientific Publishing.

Spanaki, K., Karafili, E., & Despoudi, S. (2021). AI applications of data sharing in agriculture 4.0: A framework for role-based data access control. *International Journal of Information Management*, *59*, 102350.

Sponchioni, G., Vezzoni, M., Bacchetti, A., Pavesi, M., & Renga, F. M. (2019). The 4.0 revolution in agriculture: a multi-perspective definition. In *Summer School F* (pp. 143–149). Turco-Industrial Systems Engineering.

Sridhar, A., Ponnuchamy, M., Kumar, P. S., Kapoor, A., Vo, D. V. N., & Rangasamy, G. (2023). Digitalization of the agro-food sector for achieving sustainable development goals: A review. *Sustainable Food Technology*, *1*(6), 783–802.

Stewart, R., Langer, L., Da Silva, N. R., Muchiri, E., Zaranyika, H., Erasmus, Y., Randall, N., Rafferty, S., Korth, M., Madinga, N., & De Wet, T. (2015). The effects of training, innovation and new technology on African smallholder farmers' economic outcomes and food security: A systematic review. *Campbell Systematic Reviews*, *11*(1), 1–224.

Subramanian, M., Wojtusciszyn, A., Favre, L., Boughorbel, S., Shan, J., Letaief, K. B., & Chouchane, L. (2020). Precision medicine in the era of artificial intelligence: Implications in chronic disease management. *Journal of Translational Medicine*, *18*(1), 1–12. PMID: 33298113

Sudaryanto, T, Wahida, Purba H J., Rafani, I, and Andoko E. (2022). Promoting Smart Farming based-Digital Business Technology in the Context of Agricultural Transformation in Indonesia. FFTC Journal of Agricultural Policy. https://doi.org/DOI: 10.56669/lyqu1557

Sulaiman, R., Chuluunbaatar, D., & Vishnu, S. (2018). Upscaling climate smart agriculture: lessons for extension and advisory services.

Sulaiman, R., & Aryani, H. (2021). Smart agriculture in Indonesia: Challenges and opportunities. *Indonesian Journal of Agricultural Science and Technology*, *26*(4), 112–125. DOI: 10.24831/ijast.v26i4.43021

Suprayogi, M. N., & Haryanto, B. (2023). Sustainable smart farming policy in Indonesia: A pathway to climate-resilient agriculture. *Sustainability*, *15*(3), 2149. DOI: 10.3390/su15032149

Surmaini, E., & Agus, F. (2020). CLIMATE RISK MANAGEMENT FOR SUSTAINABLE AGRICULTURE IN INDONESIA: A REVIEW / Pengelolaan Resiko Iklim untuk Pertanian Berkelanjutan di Indonesia. *Sebuah Tinjauan.*, *39*(1), 48–60. DOI: 10.21082/JP3.V39N1.2020.P48-60

Sutaryo, S., & Yuliani, K. (2023). Smart village development in Indonesia: Strengthening governance and innovation for inclusive rural growth. *Public Policy and Administration Research, 13*(2), 74–88. DOI: 10.7176/PPAR/13-2-10

Symeonaki, E., Arvanitis, K. G., Loukatos, D., & Piromalis, D. (2021). Enabling IoT wireless technologies in sustainable livestock farming toward agriculture 4.0. In *IoT- based Intelligent Modelling for Environmental and Ecological Engineering* (pp. 213–232). Springer.

Syuaib, M. F. (2016). Sustainable agriculture in Indonesia: Facts and challenges to keep growing in harmony with environment. *Agricultural Engineering International: CIGR Journal, 18*(2), 170–184.

Tao, W., Zhao, L., Wang, G., & Liang, R. (2021). Review of the internet of things communication technologies in smart agriculture and challenges. *Computers and Electronics in Agriculture, 189*, 106352.

Tatariyanto, F. (2023). Is Carbon Taxes Will Able To Tackle Forest Fires In Indonesia: A Swot Analysis In Search Of Optimal Policy. [Jurnal Ilmiah Manajemen Forkamma]. *JIMF, 6*(2), 126. DOI: 10.32493/frkm.v6i2.29298

Tibola, C. S., Silva, S. A., Dossa, A. A., & Patrício, D. I. (2018). Economically motivated food fraud and adulteration in Brazil: Incidents and alternatives to minimize occurrence. *Journal of Food Science, 83*(8), 1932–1942. DOI: 10.1111/1750-3841.14279 PMID: 30020548

Titirmare, S., Margal, P. B., Gupta, S., & Kumar, D. (2024). AI-powered predictive analytics for crop yield optimization. In *Agriculture 4.0* (pp. 89-110). CRC Press. This leads to more efficient resource use, reduced waste, and improved sustainability.

Torres-Lima, P., Canabal-Cristiani, B., & Burela-Rueda, G. (2019). Urban sustainability vs. environmental and social justice: Conflicting urban visions in Mexico City. *Cities (London, England), 94*, 77–87. DOI: 10.1016/j.cities.2019.05.031

Tugrul, B., Elfatimi, E., & Eryigit, R. (2022). Convolutional neural networks in detection of plant leaf diseases. *Revista de Agricultura (Piracicaba), 12*(8), 1192.

Tzachor, A., Devare, M., King, B., & Avin, S., & Ó hÉigeartaigh, S. (2022). Responsible artificial intelligence in agriculture requires systemic understanding of risks and externalities. *Nature Machine Intelligence, 4*(2), 104–109.

U.S. Energy Information Administration. (2021). Annual Energy Outlook 2021. https://www.eia.gov/aeo

University, I. P. B. (2024). *IPB University Lecturers Conduct Research and Visit Liberika Coffee Commodities in Podo Rukun Village*. https://www.ipb.ac.id/news

Upadhyay, A., Patel, A., Patel, A., Chandel, N. S., Chakraborty, S. K., & Bhalekar, D. G. (2025). Leveraging AI and ML in Precision Farming for Pest and Disease Management: Benefits, Challenges, and Future Prospects. *Ecologically Mediated Development: Promoting Biodiversity Conservation and Food Security*, 511-528.

Vaiyapuri, T., & Julie, G. (2025). AI and Geospatial Technologies for Climate Change Mitigation: Opportunities, Challenges, and Pathways to Sustainability. *Procedia Computer Science, 259*, 1346–1355.

Vasiloglou, M. F., Christodoulidis, S., Reber, E., Stathopoulou, T., Lu, Y., & Stanga, Z. (2020). What healthcare professionals think of "nutrition & diet" apps: An international survey. *Nutrients, 12*(8), 2214. PMID: 32722339

Vermeulen, S. J., Campbell, B. M., & Ingram, J. S. (2012). Climate change and food systems. *Annual Review of Environment and Resources, 37*(1), 195–222.

Viejo, C. G., Torrico, D. D., Dunshea, F. R., & Fuentes, S. (2019). Emerging technologies basedon artificial intelligence to assess the quality and consumer preference of beverages. *Beverages, 5*, 62. DOI: 10.3390/beverages5040062

VOSviewer Software. *Retrieved, 2025.* https://www.vosviewer.com/

Wall, P. C. (2007). Tailoring conservation agriculture to the needs of small farmers in developing countries: An analysis of issues. *Journal of Crop Improvement*, *19*(1-2), 137–155.

Wandayantolis, D., Budianta, D., Yakup, Y., & Gunawan, D. (2024). Assessing Climate-Smart Agriculture Adoption: Enhancing Rice Production Resilience in South Sumatra, Indonesia. *Journal of Smart Agriculture and Environmental Technology*, *2*(3), 93–99. DOI: 10.60105/josaet.2024.2.3.93-99

Weres, J., Mueller, W., Kozłowski, R., & Kluza, T. Advanced technologies in developing Web-based decision support systems for agriculture, in: *Proc. Of the CIGR XVIIth World Congress and the 8th World Congress on Computers in Agriculture*, 2010.

WILLIAM, B., LIANG, W., & IBRAHIM, A. (2024). Harvesting the Future: Transforming Agriculture with Climate-Smart Innovations.

Wójcik-Gront, E., Zieniuk, B., & Pawełkowicz, M. (2024). Harnessing AI-Powered Genomic Research for Sustainable Crop Improvement. *Agriculture*, *14*(12), 2299.

World Bank. (2019). *Mexico - Energy for agricultural transformation project*. World Bank Group. Retrieved from https://documents.worldbank.org/en/publication/documents-reports/documentdetail/456781562863482331/mexico-energy-for-agricultural-transformation-project

World Bank. (2021). *YESS Project: Empowering Indonesia's Rural Youth.* https://www.worldbank.org/en/results/2021/07/15/indonesia-yess-project

Wu, H. H., & Wang, S. L. (2022). Agricultural transformation under climate change: Lessons from Taiwan's innovation model. *Climate and Development Policy Review*, *8*(3), 240–256. DOI: 10.1016/j.cdpr.2022.06.005

Wu, L., Du, B., Heyden, Y. V., Chen, L., Zhao, L., Wang, M., & Xue, X. (2017). Recent advancements in detecting sugar-based adulterants in honey—A challenge. *Trends in Analytical Chemistry*, *86*, 25–38. DOI: 10.1016/j.trac.2016.10.013

Wu, T. Y., & Lin, J. H. (2021). Smart agriculture and food safety strategies under climate risk: A case study of Taiwan. *Agricultural Systems*, *193*, 103276. DOI: 10.1016/j.agsy.2021.103276

Xu, Y., Liu, X., Cao, X., Huang, C., Liu, E., Qian, S., & Zhang, J. (2021). Artificial intelligence: A powerful paradigm for scientific research. *Innovation*, *2*(4), 100179. PMID: 34877560

Xu, Z., Lv, Z., Li, J., Sun, H., & Sheng, Z. (2022). A Novel Perspective on Travel Demand Prediction Considering Natural Environmental and Socioeconomic Factors. *IEEE Intelligent Transportation Systems Magazine*.

Yadav, N., & Sidana, N. (2023). Precision Agriculture Technologies: Analysing the Use of Advanced Technologies, Such as Drones, Sensors, and GPS, In Precision Agriculture for Optimizing Resource Management, Crop Monitoring, and Yield Prediction. *Journal of Advanced Zoology, 44*.

Yadav, A. K., Gattupalli, M., Dashora, K., & Kumar, V. (2022). Key milk adulterants in India and their detection techniques: A review. *Food Analytical Methods*. Advance online publication. DOI: 10.1007/s12161-022-02427-8

Yeung, P. (2024, August 21). In Kenya, drought-weary farmers reap rewards of early warnings. Context News. Retrieved July 9, 2025, from https://www.context.news/climate-risks/in-kenya-drought-weary-farmers-reap-rewards-of-early-warnings

Yu, Q., Zhang, M., Mujumdar, A. S., & Li, J. (2024). AI-based additive manufacturing for futurefood: Potential applications, challenges and possible solutions. *Innovative Food Science & Emerging Technologies*, *92*, 103599. DOI: 10.1016/j.ifset.2024.103599

Yusuf, I. S., Yusuf, M. O., Panford-Quainoo, K., & Pretorius, A. (2024). A geospatial approach to predicting desert locust breeding grounds in Africa. *arXiv preprint arXiv:2403.06860.*

Yusuf, A. A., & Francisco, H. (2010). *Climate change vulnerability mapping for Southeast Asia.* Economy and Environment Program for Southeast Asia.

Zepeda, L., & Deal, D. (2008). Think before you eat: Photographic food diaries as intervention tools to change dietary decision-making and attitudes. *International Journal of Consumer Studies, 32*(6), 692–698.

Zhang, J., Zhang, X., Dediu, L., & Victor, C. (2011). Review of the current application of fingerprinting allowing detection of food adulteration and fraud in China. *Food Control, 22*(8), 1126–1135. DOI: 10.1016/j.foodcont.2011.01.019

Zheng, Z. Y., Guo, X. N., Zhu, K. X., Peng, W., & Zhou, H. M. (2017). Artificial neural network-genetic algorithm to optimize wheat germ fermentation condition: Application to the production of two anti-tumor benzoquinones. *Food Chemistry, 227*, 264–270. PMID: 28274431

Zhu, K., Lammers, K., Zhang, K., Arunachalam, C., Bhattacharya, S., Li, J., . . . Li, Z. (2025). Advancement and Field Evaluation of a Dual-arm Apple Harvesting Robot. *arXiv preprint arXiv:2506.05714.*

Zuo, M., Ji, H., Su, L., Zhang, Y., Yan, W., & Zhang, Q.. (2024). Advancements andmodern applications of artificial intelligence in ensuring food safety. *J Chin Inst Food Sci Technol, 24*, 1–13.

About the Contributors

Eduard Babulak is an accomplished international scholar, researcher, consultant, educator, professional engineer, and polyglot. He successfully published his research which has been cited by scholars from all over the world. He is a Fellow of Royal Society RSA and Chartered Fellow, Mentor, and Elite Group Member of the British Computer Society in the UK.

***

Kashvi Chaturvedi, is studying B.Tech. specialisation in Artificial Intelligence and Data Science at Ajeenkya D Y Patil University. She started solidifying her programming skills by completing various certification courses like Data Structures in C, Tableau functions and Operating Systems along with doing few projects using languages like Python, C, C++ and creating Web Page using HTML, CSS and Java Script. She had also attended an Arduino-Based Automation course at Dolphin Labs to enhance her software knowledge. There, she learned how to work with sensors, microcontrollers, and circuit design. She has also written multiple review papers on topics such as Geographic Information Systems, Comparative Analysis of BJT Configurations, Swiss Bank vs. European Central Bank: Data Analysis. She is also researching on and learning many topics such as Small Modular Reactors, Bibliometric Analysis, UNSDGs etc. Because of her interest in latest technologies, she also participated in Global Hyperloop Competition hosted by IIT Madras representing her college as team Vac-Vectoor where she learnt more about various subtopics from the mentors.

Csongor Czipf, Ph.D. Honorary Associate Professor, Assistant Professor Széchenyi István University, Győr, Hungary Gyula Kautz Faculty of Economics, Department of International Applied Economics Ongoing teaching experience at Széchenyi István University since 1999. Appointed as Regional Economic Manager of the Piarist Order Hungarian Province in Budapest as of January 2022. Master's degree in Religious Education and Dormitory Teaching obtained from the College of Theology in Győr in 2017. Doctoral degree awarded in 2011 by Janus Pannonius University of Pécs. Research specialization: corporate evaluation from a human resources perspective. Bachelor's degree in Economics with a specialization in Comparative Economic Policy completed in 1999 at Széchenyi István College.

Susanta Das is a seasoned educator with a proven record of teaching and mentoring diverse students, working at multiple institutions, and providing service to the various initiatives of organizations. He received his Ph.D. and M.A. degrees from Western Michigan University (WMU), USA, and M.Sc. degree from Banaras Hindu University (BHU), India, all in Physics. He continued his research as a Marie-Curie post-doctoral fellow at Stockholm University, Stockholm, Sweden on beam diagnostics for the DESIREE (Double ElectroStatic Ion Ring ExpEriment) in the project DITANET (Diagnostic Techniques for particle Accelerators – a Marie-Curie initial training NETwork), at the Indian Institute of Science Education and Research-Kolkata, India on high-pressure physics as a Project Scientist-B, and at the University of Electro-Communications, Tokyo, Japan on ion-surface interactions as a Post-doctoral fellow. Throughout his career, Dr. Das worked and collaborated with many researchers, Ph.D., master, and visiting students from different countries. He further visited many countries to discuss research and an international conference participant (UK, Italy, Germany, Greece, Belgium, Bulgaria, Romania, Brazil etc.). He worked at several institutions before assuming his current role at Ajeenkya D Y Patil University. He received the Marie-Curie post-doctoral fellowship in Sweden, Science Academies' Summer Research Fellowship in India, Gwen Frostic Doctoral Fellowship, Department Graduate Research and Creative Scholar Award by WMU, and Leo R. Parpart Doctoral Fellowship by Dept. of Physics, WMU, among many others, throughout his academic journey. He co-authored several research articles in WoS/Scopus indexed international journals, conference proceedings, and scholarly book chapters and co-edited several books published by renowned international publishing houses (IGI-Global, Springer Nature, Nova Science Publishers, Willey, & Scrivener-Willey, Notion Press). At present, Dr. Das continues to delve into cutting-edge fields, with a keen interest in nanotechnology, UN SDGs, and data science. His multifaceted contributions, spanning teaching, research, and administrative leadership, showcase a dedicated professional who is instrumental in advancing the vision and mission of the institutions he serves.

Pooja Dehankar is working in Department of Artificial intelligence and Data Science at Ajeenkya D Y Patil School of Engineering, affiliated to Savitribai Phule Pune University, Pune. She has completed BE in Computer Engineering from Priyadarshini College of Engineering, Nagpur, and ME in Information Technology from Sinhgad College of Engineering, Pune. She has done Post Graduate Diploma in Advance Computing. She has 17 years of experience in academics. Her areas of interest are Artificial Intelligence, Cyber Security, and Data Mining. She has published papers in Scopus Journals. She actively participates in workshops, faculty development programs, short-term training programs, conferences, and webinars. She is pursuing Ph.D. from Symbiosis Center for Research and Innovation, Symbiosis International (Deemed University), Pune.

H. Vidyalakshmi is affiliated with the Department of Nutrition and Dietetics at the School of Life Sciences, Ooty Campus, JSS Academy of Higher Education and Research, Mysuru. Based in Ooty, Tamil Nadu, her research interests span food biotechnology, sports nutrition, nutraceuticals, artificial intelligence in healthcare, and public health nutrition.

Raj Kumar is a distinguished faculty member at the Institute of Management Studies and Research, Maharshi Dayanand University, located in Rohtak, Haryana (Postal Code: 124001). With a strong academic presence and a commitment to excellence in management education, Prof. Kumar has significantly contributed to the academic and professional development of students in the field of management studies.

S. N Kumar is currently an Associate. Professor, Dept. of Electrical and Electronics Engineering., Amal Jyothi College of Engineering, Kanjirappally, Kerala. He was awarded PhD in Electronics and Communication Engineering from Sathyabama Institute of Science and Technology, Chennai. He has 15 years of teaching experience, and his areas of interest include Medical Image Processing and Embedded System Applications in Telemedicine. He was the Co- Principal Investigator of the DST IDP-funded project and research fellow of the RCA scheme, NTU Singapore. He has edited 3 books, authored 32 book chapters and 30 publications in peer reviewed journals. He had authored 12 text books in engineering discipline.

T.C. Manjunath was born on 6th February 1967. He did his BE from RV College of Engg. in 1989, ME in Robotics & Automation from LD College of Engg. in 1992 with First Rank and PhD in Systems & Control Engg. from IIT Bombay in the year 2007. He is currently working as the Dean Research R D and as Professor in the Dept. of Computer Science & Engineering with specialization in IoT, Cyber Security & Blockchain Technology in Rajarajeswari College of Engineering, Bengaluru, Karnataka. He has completed B.E. Degree from R.V. College of Engg., M.E. Govt.'s LD College of Engg., Gujarat Univ., Ahmedabad in First Rank and Ph.D. in Systems and Control Engineering from the Indian Institute of Technology Bombay IIT Bombay He has got a teaching academic, research and administrative experience of more than 35+ years 10 years as Principal, 8 years as HoD rest as faculty and has published more than 1200+ research papers in various national & international journals and conferences, with 175+ patents, 25 book-chapters, 25 awards and also published 25 text books in various field of science, engineering and technology, completed projects worth Rs. 9 Crores till date apart from guiding 10 PhD scholars. He is a member of 30 Professional Society Chapters and is a fellow of IE, IETE, AMIEE and a chartered engineer. He got the best research scholar award in 2006 in IIT Bombay amongst all the research scholars of the institute. He has got 10 awards in academics and research activities.

P. Selvakumar is currently an Associate Professor in the Department of Science and Humanities at Nehru Institute of Technology, Coimbatore,

Tamil Nadu, India. Prior to joining Nehru Institute of Technology, he held positions as an Assistant Professor in the Department of Science and Humanities various engineering colleges in Coimbatore, Tamil Nadu, India, spanning from 2011 to 2023. Dr.P. Selvakumar's academic and research interests lie primarily in the field of phytochemistry. He has made significant contributions to his field, with an impressive record that includes 30 journal publications, 12 patents, and 20 conference papers. Additionally, he has organized 5 seminars and conferences, further demonstrating his commitment to academic and scientific exchange. Throughout his career, Dr. P.Selvakumar has been recognized for his dedication and excellence in teaching, having been honored with several awards, including the Best Teacher Awards. He holds a prominent position in the academic community, contributing actively to research.

R. Josphineleela has received her Ph.D (Computer Science Engineering) from Sathyabama University, India in 2013. She has completed her M. E (computer science and Engineering) in Sathyabama University. She has more than 23 years' experience in the field of Computer Science and she is currently working as a Professor in the department of computer Science and engineering at Panimalar Engineering College. She has published more than 50 papers at national and international level. Her research is in the field of Image processing, Neural Network, Artificial Intelligence, Biomedical Imaging and Soft Computing etc. She has received a Distinguished Professor award from Computer Society of India and received Best Project award from Dr.Kalam Educational Trust for Tribal University, Best Teacher Award from IEEE.She got Best Paper award from Computer Society of India,she has received "Certificate of Appreciation" for contributing as a Proctor in "IEEEXtreme 12.0" Programming. She has received "In appreciation for fostering an ecosystem bridging Government, Industry and Academia award" from "India Innovation challenge design contest 2018 from DST & Texas Instrument". She has authored Python, IT Essentials, Computer Networks and Computational Intelligence

Ragupathi T received his B.E. degree in Computer Science and Engineering from Anna University, Tamil Nadu, in the year 2009 and the M.E. Master's degree in Computer Science and Engineering from

Anna University, Tamil Nadu in the year 2011. He completed his Doctor of Philosophy in Computer Science and Engineering from Annamalai University, Chidambaram, Tamil Nadu, in the year 2023.He has more than 13.5 years of teaching experience. His research area includes Deep Learning, Machine Learning, IoT and web technology. He has published papers in several International Journals, international and national conferences

V.Vaissnave is obtained B.Tech Information Technology in PSR Engg college, Anna University. She obtained her Master of Engg in Anand Institute of Science and Technology, Anna University. And she obtained her Full time Research in the Department of Computer Science and Engineering, School of Computing, Kalasalingam Academy of Research and Education, Krishnankoil, Tamilnadu. Her area of interest is Machine learning, Deeplearning, and Natural Language Techniques.

Asamoah Oppong Zadok is the Founder of Sustaina Harvest, a mission-driven agribusiness based in Akim Oda – Adjobue, Eastern Region, Ghana, that integrates eco-friendly crop and livestock production with social impact initiatives focused on empowering women, youth, and persons with disabilities. He is a researcher and agribusiness professional with expertise in sustainable agriculture, agribusiness, and agricultural policy, with research interests that include consumer demand, local food systems, and the use of AI in agriculture to improve productivity, decision-making, and sustainability. Asamoah holds a Master of Science in Sustainable Agriculture from Lincoln University, USA, and a Bachelor of Science in Agribusiness from the University of Cape Coast, Ghana. He has presented his research at international conferences and serves as a peer reviewer for academic journals in the fields of economics and agriculture.

Index

G

H

I

M

P

R

S

T

www.ingramcontent.com/pod-product-compliance
Lightning Source LLC
LaVergne TN
LVHW080855300926
849162LV00003B/427